Nuclear Power Systems: Principles and Applications for Engineers

Holbein

Copyright © [2023]

Author: Holbein

Title: Nuclear Power Systems: Principles and Applications for Engineers

This book is a self-published work by the author Holbein

ISBN:

TABLE OF CONTENTS

Chapter 1: Introduction to Nuclear Power Systems

Overview of Nuclear Energy Engineering

Nuclear energy engineering plays a crucial role in meeting the growing global demand for clean and reliable energy. In this subchapter, we will provide engineers specializing in energy engineering with a comprehensive overview of the principles and applications of nuclear power systems. By understanding the fundamentals of nuclear energy engineering, engineers will be equipped to contribute to the development and operation of safe, efficient, and sustainable nuclear power plants.

The subchapter begins by introducing the basic principles of nuclear physics, including the structure of the atom, nuclear reactions, and the concept of radioactivity. These fundamental concepts are essential for engineers to grasp the underlying principles of nuclear energy generation.

Next, we delve into the various types of nuclear reactors, including pressurized water reactors (PWRs), boiling water reactors (BWRs), and advanced reactor designs. Engineers will gain an understanding of the core components and processes involved in nuclear reactors, such as fuel rods, control rods, coolant systems, and steam generators. Safety features and protocols, including emergency shutdown systems and containment structures, are also covered in detail.

The subchapter also explores the fuel cycle of nuclear power systems, from uranium mining and enrichment to spent fuel management and disposal. Engineers will learn about the challenges and opportunities associated with each stage of the fuel cycle and the importance of ensuring the safe handling and storage of radioactive materials.

Furthermore, we discuss the environmental impact of nuclear power and its role in mitigating climate change. Engineers will gain insights into the carbon footprint of nuclear energy compared to fossil fuels, as well as the potential for advanced nuclear technologies to further reduce environmental impacts.

Lastly, the subchapter addresses the socio-economic aspects of nuclear energy engineering. Engineers will learn about the economic viability of nuclear power plants, the regulatory framework governing their operation, and the social acceptance of nuclear energy within different communities.

Throughout this subchapter, engineers specializing in energy engineering will gain a comprehensive understanding of nuclear power systems, enabling them to contribute to the design, operation, and advancement of this critical source of clean, reliable, and sustainable energy.

Historical Development of Nuclear Power Systems

Introduction:
The historical development of nuclear power systems has played a significant role in shaping the field of energy engineering. Engineers have been at the forefront of this journey, constantly innovating and improving upon the principles and applications of nuclear power. This subchapter aims to provide engineers, particularly those specializing in energy engineering, with a comprehensive overview of the historical development of nuclear power systems.

Early Discoveries and Milestones:
The story of nuclear power systems begins with the discovery of radioactivity by Antoine Henri Becquerel in 1896, followed by Marie and Pierre Curie's significant contributions to the understanding of radioactive elements. These discoveries laid the foundation for nuclear science and technology. In the 1930s, scientists like Enrico Fermi and Hans Bethe made breakthroughs in nuclear fission, leading to the development of the first-ever controlled nuclear chain reaction in 1942 at the University of Chicago.

The Manhattan Project and Commercialization:
The Manhattan Project, initiated during World War II, aimed at developing atomic weapons. However, it also paved the way for the commercialization of nuclear power. The successful construction and operation of the first nuclear reactor, the Chicago Pile-1, marked a significant milestone in nuclear power history. After the war, research efforts focused on harnessing nuclear energy for peaceful purposes, leading to the establishment of the first commercial nuclear power plant, the Shippingport Atomic Power Station, in 1957.

Advancements and Expansions:
Throughout the 1960s and 1970s, there was a rapid expansion of nuclear power systems worldwide. Engineers played a crucial role in improving reactor designs, safety measures, and operational efficiency. The development of pressurized water reactors (PWRs) and boiling water reactors (BWRs) became the dominant technologies for commercial nuclear power plants. Additionally, advancements were made in fuel technology, waste management, and radiation protection.

Challenges and Aftermath:
The nuclear power industry faced several challenges in the late 20th century, including safety concerns, public perception, and the high cost of construction and operation. The Three Mile Island accident in 1979 and the Chernobyl disaster in 1986 significantly impacted the industry, leading to increased safety regulations and a decline in nuclear power plant construction. However, engineers continued to work towards enhancing safety features, such as passive cooling systems and standardized designs.

Modern Developments:
In recent years, engineers have been exploring advanced reactor designs, including small modular reactors (SMRs), thorium-based reactors, and fusion reactors. These developments aim to address the challenges faced by traditional nuclear power systems, such as waste disposal and non-proliferation concerns, while providing sustainable and clean energy solutions for the future.

Conclusion:
The historical development of nuclear power systems has been a journey of innovation, challenges, and advancements. Engineers in the

field of energy engineering have played a vital role in shaping the industry and addressing the concerns associated with nuclear power. By understanding the historical context, engineers can build upon past achievements and continue to drive the development of safe, efficient, and sustainable nuclear power systems.

Advantages and Disadvantages of Nuclear Power

Introduction:
Nuclear power is a significant component of the global energy mix and plays a crucial role in meeting the growing energy demands of our modern world. As engineers in the field of energy engineering, understanding the advantages and disadvantages of nuclear power is essential for designing, operating, and evaluating nuclear power systems. This subchapter aims to provide you with a comprehensive overview of the benefits and drawbacks associated with nuclear power.

Advantages of Nuclear Power:
1. Low Greenhouse Gas Emissions: Nuclear power plants produce electricity without emitting large amounts of greenhouse gases, such as carbon dioxide. This characteristic makes nuclear power an attractive option for reducing the carbon footprint and combating climate change.

2. High Energy Density: Nuclear power has an incredibly high energy density, meaning a small amount of nuclear fuel can generate a substantial amount of electricity. This efficiency allows for more energy production with fewer resources.

3. Base Load Power: Nuclear power plants provide a stable and reliable source of continuous electricity, making them suitable for meeting base load power demands. Unlike renewable energy sources like solar and wind, nuclear power can operate consistently, ensuring a constant supply of electricity.

4. Fuel Availability: Uranium, the primary fuel used in nuclear power plants, is abundant in many regions globally. This availability ensures

a long-term and sustainable energy source, reducing dependency on fossil fuels.

Disadvantages of Nuclear Power:
1. Radioactive Waste: The generation of radioactive waste is one of the significant concerns associated with nuclear power. Proper management and disposal of this waste are crucial to prevent any adverse environmental and health effects.

2. Cost and Complexity: Building and maintaining nuclear power plants require significant capital investment and specialized expertise. The complexity of nuclear technology adds to the overall cost, making it a financially challenging option for some countries.

3. Safety Risks: While nuclear power plants have multiple safety mechanisms in place, accidents like the Chernobyl and Fukushima incidents highlight the potential risks associated with nuclear power. Ensuring the highest safety standards and implementing robust regulatory frameworks are essential to mitigate these risks effectively.

4. Nuclear Proliferation: The use of nuclear technology for power generation can potentially contribute to the proliferation of nuclear weapons. Strict international agreements and stringent control measures are necessary to prevent the misuse of nuclear materials and technologies.

Conclusion:
Understanding the advantages and disadvantages of nuclear power is crucial for engineers in the field of energy engineering. While nuclear power offers numerous benefits such as low greenhouse gas emissions, high energy density, and fuel availability, it also presents challenges

related to radioactive waste, cost, safety risks, and proliferation. By critically evaluating these factors and implementing appropriate measures, engineers can contribute to the development and utilization of nuclear power in a safe, sustainable, and responsible manner.

Chapter 2: Fundamentals of Nuclear Physics

Atomic Structure and Nuclear Reactions

In this subchapter, we will delve into the fascinating world of atomic structure and nuclear reactions, exploring the fundamental principles that underpin nuclear power systems. As engineers specializing in energy engineering, it is essential to have a solid understanding of the atomic structure and the intriguing phenomena that occur at the nuclear level.

At the heart of nuclear power systems lies the atom, the building block of matter. We will begin by revisiting the structure of the atom, understanding its various components, such as protons, neutrons, and electrons, and how they interact to form stable and unstable nuclei.

Next, we will explore the concept of isotopes and the significance of their stability or instability in nuclear reactions. We will discuss the role of the strong and weak nuclear forces in determining the stability of nuclei and the implications for nuclear power systems.

Moving on, we will dive into the world of nuclear reactions, understanding the processes that occur when nuclei interact with one another. We will explore both fusion and fission reactions, discussing their differences, advantages, and challenges. Engineers in the field of energy engineering must be familiar with the intricacies of these reactions to optimize the performance and safety of nuclear power systems.

Furthermore, we will examine the concept of nuclear decay, where unstable isotopes undergo spontaneous transformations to achieve a more stable state. We will explore the different types of radioactive decay, including alpha, beta, and gamma decay, and their associated energy releases. Understanding these decay processes is crucial for engineers involved in the design and management of nuclear power systems.

To conclude the subchapter, we will touch upon the concepts of nuclear cross-sections, reaction rates, and neutron interactions. These parameters play a vital role in the analysis and prediction of nuclear reactions, enabling engineers to optimize the efficiency and safety of nuclear power systems.

By the end of this subchapter, engineers specializing in energy engineering will have a comprehensive understanding of atomic structure and nuclear reactions. Armed with this knowledge, they will be well-equipped to design, analyze, and optimize nuclear power systems, ensuring sustainable and reliable energy generation for the future.

Note: The content provided here is a brief summary of the subchapter "Atomic Structure and Nuclear Reactions" from the book "Nuclear Power Systems: Principles and Applications for Engineers". The actual content in the book may be more detailed and comprehensive.

Radioactivity and Radiation

In the field of energy engineering, it is essential to have a thorough understanding of radioactivity and radiation, as they play a crucial role in nuclear power systems. This subchapter aims to provide engineers with a comprehensive overview of these concepts, their properties, and their applications in the field of nuclear power.

Radioactivity refers to the spontaneous emission of radiation from the nucleus of an atom. This emission occurs due to the instability of certain atomic nuclei, known as radionuclides. Radionuclides can emit different types of radiation, including alpha particles, beta particles, and gamma rays. Each type of radiation possesses distinct characteristics and poses varying levels of risk to human health and the environment.

Alpha particles are positively charged particles consisting of two protons and two neutrons, similar to a helium nucleus. Due to their large size and positive charge, alpha particles have a limited range and can be easily stopped by a sheet of paper or a few centimeters of air. Beta particles, on the other hand, are high-speed electrons or positrons emitted by certain radionuclides. They have a smaller mass and can penetrate further than alpha particles but can be stopped by a few millimeters of aluminum or plastic.

Gamma rays are electromagnetic radiation of high energy and frequency. Unlike alpha and beta particles, gamma rays have no mass or charge, enabling them to penetrate several centimeters of lead or even meters of concrete. Due to their penetrating ability, gamma rays pose the greatest risk to human health and require appropriate shielding measures.

Understanding the properties and behavior of radiation is of paramount importance when designing and operating nuclear power systems. Engineers must ensure the safety of both workers and the general public by implementing robust shielding measures, monitoring radiation levels, and minimizing exposure.

Furthermore, engineers need to be familiar with the various applications of radiation in energy engineering. Nuclear power systems rely on controlled nuclear reactions to generate heat, which is then converted into electricity. Radioactive isotopes are used in the production of nuclear fuels, such as uranium and plutonium, while radiation is employed in the inspection and testing of materials, such as welds and pipelines, to ensure their integrity and reliability.

In conclusion, this subchapter has provided engineers specializing in energy engineering with a comprehensive understanding of radioactivity and radiation. By grasping the properties, risks, and applications of radiation, engineers can contribute to the safe and efficient operation of nuclear power systems, ensuring a reliable and sustainable energy source for the future.

Nuclear Fission and Fusion

In the realm of energy engineering, nuclear power systems have emerged as a groundbreaking technology, offering immense potential for generating clean and efficient electricity. Central to these systems are the processes of nuclear fission and fusion, which harness the power of atomic reactions to produce a sustainable source of energy.

Nuclear fission refers to the splitting of an atomic nucleus into two or more smaller nuclei, often accompanied by the release of a large amount of energy. This process is achieved by bombarding a heavy nucleus, such as uranium-235 or plutonium-239, with a neutron, causing it to become unstable and split into two smaller nuclei, along with the release of additional neutrons and vast amounts of energy. These released neutrons can then trigger a chain reaction, sustaining the fission process and generating a substantial amount of heat.

The heat generated by nuclear fission is then used to produce steam, which drives a turbine connected to an electrical generator, ultimately producing electricity. One of the key advantages of nuclear fission is its ability to generate an enormous amount of energy from a small amount of fuel, making it highly efficient and cost-effective. Additionally, nuclear power systems do not emit greenhouse gases, making them an environmentally friendly alternative to traditional fossil fuel-based power plants.

On the other hand, nuclear fusion involves the merging of two atomic nuclei to form a heavier nucleus, accompanied by the release of a tremendous amount of energy. Fusion reactions occur at extremely high temperatures and pressures, similar to those found in the core of stars. Although fusion has been pursued as a potential energy source for several decades, scientists are yet to achieve commercial-scale fusion reactors due to the technical challenges involved in confining and controlling the extremely hot plasma.

However, the potential benefits of nuclear fusion are significant. Fusion reactions, unlike fission, do not produce long-lived radioactive waste, and the fuel required for fusion, such as isotopes of hydrogen, is abundant in nature. If successfully harnessed, fusion could provide an almost limitless supply of clean energy, free from the concerns associated with nuclear waste and the risks of meltdowns.

As engineers in the field of energy engineering, understanding the principles and applications of nuclear fission and fusion is crucial. These processes hold the key to unlocking the true potential of nuclear power systems, offering a sustainable and efficient source of energy that can meet the growing global demand while reducing our reliance on fossil fuels. By delving deeper into the intricacies of nuclear fission and fusion, engineers can contribute to the development of safer and more advanced nuclear power technologies, shaping the future of energy generation.

Chain Reactions and Criticality

In the realm of nuclear power systems, the concept of chain reactions and criticality lies at the very heart of their operation. This subchapter aims to provide engineers, specifically those in the field of energy engineering, with a comprehensive understanding of these fundamental principles. By delving into the intricacies of chain reactions and criticality, engineers can effectively design, operate, and maintain nuclear power systems, ensuring the safe and efficient generation of energy.

A chain reaction is a self-sustaining process in which the fission of one atomic nucleus produces neutrons that, in turn, cause the fission of other atomic nuclei, creating a cascade effect. This continuous release of energy is the basis of nuclear power generation. Understanding the characteristics of a chain reaction is crucial for engineers, as it enables them to control and regulate the nuclear reactions occurring within a reactor.

Criticality, on the other hand, refers to the condition in which a nuclear reactor is able to sustain a self-sustaining chain reaction. Achieving criticality involves maintaining a delicate balance between the number of neutrons produced and lost within a reactor. Engineers must carefully monitor parameters such as reactor geometry, fuel composition, and neutron moderation to ensure that the reactor remains in a critical state.

The study of chain reactions and criticality involves various mathematical models and calculations. Engineers must be well-versed in these techniques to accurately predict and analyze the behavior of nuclear power systems. Through the use of diffusion theory, transport theory, and Monte Carlo methods, engineers can simulate neutron transport, reaction rates, and power distributions within a reactor. These simulations aid in the design and optimization of reactor cores, fuel assemblies, and control mechanisms.

Moreover, engineers must also consider the concept of subcriticality, which refers to a condition where the number of neutrons produced is insufficient to sustain a self-sustaining chain reaction. Subcriticality plays a significant role in reactor safety, as it allows for the control and shutdown of a reactor without the risk of a runaway chain reaction.

In conclusion, chain reactions and criticality are pivotal concepts in the field of nuclear power systems, particularly for engineers specializing in energy engineering. A thorough understanding of these principles enables engineers to design, operate, and maintain nuclear power systems that are safe, efficient, and reliable. By continuously advancing their knowledge in this area, engineers can contribute to the sustainable development and utilization of nuclear energy for the betterment of society.

Chapter 3: Nuclear Reactor Design and Components

Types of Nuclear Reactors

In the world of energy engineering, nuclear power systems play a crucial role in meeting the increasing demand for clean and sustainable energy sources. Nuclear reactors, the heart of these power systems, are designed to harness the immense power of nuclear fission to generate electricity. This subchapter will delve into the various types of nuclear reactors, shedding light on their unique features and applications.

1. Pressurized Water Reactors (PWRs): PWRs are the most common type of nuclear reactors used worldwide. They utilize enriched uranium as fuel and employ pressurized water as both the coolant and moderator. PWRs are known for their high thermal efficiency and excellent safety features, making them ideal for large-scale electricity generation.

2. Boiling Water Reactors (BWRs): BWRs are similar to PWRs but differ in the way they handle coolant. In BWRs, water is allowed to boil directly in the reactor core, producing steam that drives the turbine. These reactors are simpler in design and require less complex plumbing, resulting in reduced capital costs.

3. Heavy Water Reactors (HWRs): HWRs use heavy water (deuterium oxide) as both the coolant and moderator. They offer advantages such as the ability to use natural uranium as fuel, reducing the need for uranium enrichment. HWRs are desirable for their high neutron

economy and the potential for breeding new fuel, making them an attractive option for countries with limited uranium resources.

4. Gas-Cooled Reactors (GCRs): GCRs employ gases such as carbon dioxide or helium as coolants. These reactors have the advantage of high-temperature operation, allowing for more efficient electricity generation. GCRs are also well-suited for cogeneration applications, where the excess heat produced during electricity generation is utilized for other purposes, such as district heating.

5. Sodium-Cooled Fast Reactors (SFRs): SFRs utilize liquid sodium as both the coolant and moderator. They operate at higher temperatures and have the ability to efficiently convert fertile material into usable fuel, effectively extending the fuel cycle. SFRs are considered a promising option for future advanced nuclear power systems due to their high efficiency and potential for reducing nuclear waste.

Understanding the different types of nuclear reactors is crucial for engineers involved in energy engineering. Each reactor type has its own advantages and limitations, and choosing the appropriate design depends on various factors, including safety, cost, availability of fuel, and specific energy requirements. By exploring these reactor types, engineers can contribute to the development and optimization of nuclear power systems that meet the ever-growing energy demands while ensuring a sustainable and secure future.

Reactor Core Design and Fuel Assembly

In the realm of energy engineering, nuclear power has emerged as a promising and efficient alternative to traditional fossil fuel sources. Nuclear power systems rely on the ingenious design of reactor cores and fuel assemblies to harness the immense energy potential of nuclear fission reactions. This subchapter aims to provide engineers with a comprehensive understanding of reactor core design and fuel assembly.

At the heart of a nuclear power system lies the reactor core, which houses the fuel assemblies. The reactor core serves as the central component where controlled nuclear fission reactions take place, generating heat that is ultimately converted into electricity. Efficient reactor core design is crucial to ensure optimal performance, safety, and longevity of the nuclear power system.

The core design involves various considerations, including fuel type, geometry, and arrangement. Engineers must carefully select the fuel type, such as uranium or plutonium, based on factors like availability, cost, and fuel cycle efficiency. The geometric arrangement of fuel assemblies within the core is critical for maintaining a sustainable chain reaction while managing neutron flux, power distribution, and coolant flow.

Fuel assemblies, comprising fuel rods, control rods, and other components, play a vital role in maintaining and controlling the nuclear reactions. Fuel rods contain the fissile material, which undergoes fission and releases energy. Control rods, made of neutron-absorbing materials like boron or cadmium, help regulate the reaction by absorbing excess neutrons, thus controlling the power output. Engineers must design fuel assemblies with a balance between maximizing energy production and ensuring safety and stability.

Furthermore, the subchapter will delve into the core cooling and moderation systems. Efficient cooling is essential to prevent overheating and ensure the integrity of the fuel. Engineers must design effective coolant systems, such as water or liquid metal cooling, that can handle the immense heat generated within the core.

The subchapter will also discuss the importance of neutron moderation, which involves slowing down fast neutrons to maintain a sustained chain reaction. Various moderators, including light water, heavy water, or graphite, can be employed based on their moderation properties.

In summary, reactor core design and fuel assembly are crucial aspects of nuclear power systems. Engineers in the field of energy engineering must have a strong grasp of these concepts to design and optimize safe, efficient, and reliable nuclear power plants. By understanding the principles and considerations behind reactor core design and fuel assembly, engineers can contribute to the advancement and implementation of nuclear power systems as a sustainable energy source for the future.

Moderator and Coolant Systems

In the realm of nuclear power systems, the moderator and coolant systems play vital roles in ensuring safe and efficient operation. These systems are of utmost importance in maintaining the controlled chain reaction and transferring heat away from the reactor core. This subchapter aims to provide engineers in the field of Energy Engineering with a comprehensive understanding of these systems, their functions, and their significance in nuclear power plants.

The moderator system is responsible for slowing down the fast neutrons produced during nuclear fission. By reducing their kinetic energy, the moderator ensures that the neutrons are more likely to induce further fission events, sustaining the chain reaction. Common moderators include light water (H_2O), heavy water (D_2O), and graphite. Each moderator possesses its own advantages and disadvantages, which engineers must carefully consider when designing nuclear power systems.

Coolant systems, on the other hand, are responsible for transferring the heat generated in the reactor core to produce steam, which in turn drives turbines to generate electricity. These systems ensure that the core remains at a stable temperature and prevent any overheating that could lead to a catastrophic failure. Common coolants used in nuclear power plants include water, gas, and liquid metals. The choice of coolant depends on factors such as thermal efficiency, heat transfer properties, and the specific requirements of the power plant.

Engineers must consider numerous factors when designing moderator and coolant systems. These include the physical and chemical properties of the materials used, the desired power output, safety considerations, and the overall efficiency of the system. Additionally, engineers must also consider the impact of these systems on the environment, ensuring that they are sustainable and do not pose any significant risks to ecosystems or human health.

Understanding the intricacies of moderator and coolant systems is crucial for engineers working in the field of Energy Engineering. By comprehending the principles behind these systems, engineers can make informed decisions during the design, construction, and operation of nuclear power plants. Moreover, an in-depth knowledge of these systems enables engineers to troubleshoot any issues that may arise and ensure the safe and reliable operation of nuclear power plants.

In conclusion, the moderator and coolant systems are essential components of nuclear power plants. They play a significant role in maintaining controlled chain reactions, transferring heat from the reactor core, and generating electricity. Engineers specializing in Energy Engineering must possess a deep understanding of these systems to ensure the safe, efficient, and sustainable operation of nuclear power plants.

Control and Safety Systems

In the field of energy engineering, control and safety systems play a crucial role in ensuring the safe and efficient operation of nuclear power systems. These systems are designed to monitor, control, and protect various components of a nuclear power plant, including the reactor, turbines, and auxiliary systems.

The primary objective of control systems in a nuclear power plant is to maintain stable and safe operation. This involves regulating reactor power, controlling reactor coolant flow, and adjusting control rod positions to maintain the desired power level. To achieve this, a combination of sensors, actuators, and control algorithms are employed. The control system continuously monitors various parameters such as neutron flux, coolant temperature, and pressure to make necessary adjustments and ensure the stability of the reactor.

Safety systems, on the other hand, are designed to mitigate potential risks and prevent accidents. These systems are divided into two categories: active and passive safety systems. Active safety systems rely on external sources of power, such as electricity, to function. They include emergency core cooling systems, which provide additional cooling to the reactor in case of an accident, and containment systems, which prevent the release of radioactive materials into the environment. Passive safety systems, on the other hand, operate without external power sources and rely on natural forces such as gravity and convection for their function. Examples of passive safety systems include passive heat removal systems and passive containment cooling systems.

In addition to control and safety systems, nuclear power plants are equipped with various other safety measures, such as redundant systems, diverse control mechanisms, and rigorous safety protocols. These measures are implemented to ensure that any potential failure or anomaly is detected and addressed promptly.

The engineers involved in the design and operation of nuclear power systems must have a deep understanding of control and safety systems. They need to possess knowledge of control theory, instrumentation, and safety regulations specific to nuclear energy. Additionally, they must be skilled in the use of simulation tools and have practical experience in troubleshooting and maintenance of control and safety systems.

In conclusion, control and safety systems are vital components of nuclear power systems. They are responsible for the stable and safe operation of the plants. Engineers specializing in energy engineering play a critical role in designing, implementing, and maintaining these systems to ensure the efficient and secure generation of nuclear power.

Chapter 4: Nuclear Fuel Cycle

Mining and Processing of Uranium

Uranium is a crucial element in the production of nuclear power, serving as the primary fuel source for nuclear reactors. The mining and processing of uranium are essential steps in harnessing this valuable resource. In this subchapter, we will explore the various stages involved in the extraction and refinement of uranium, providing engineers in the field of energy engineering with a comprehensive understanding of these processes.

The mining of uranium begins with the identification and selection of suitable deposits. Geological surveys and exploration programs play a vital role in locating potential uranium-rich sites. Once identified, the mining process involves extracting uranium ore from the ground. This can be achieved through open-pit or underground mining methods, depending on the specific characteristics of the deposit.

Once the uranium ore is obtained, it undergoes a series of processing steps to extract the valuable uranium content. The initial stage involves crushing and grinding the ore into smaller particles, facilitating the release of uranium from the surrounding rock. This step is followed by a process called leaching, where a chemical solution is used to dissolve the uranium from the crushed ore.

After leaching, the uranium solution is purified through a process known as solvent extraction. This technique involves separating the uranium from other elements and impurities present in the solution.

The purified uranium is then precipitated and dried, resulting in a concentrated uranium powder called yellowcake.

Yellowcake serves as the feedstock for the subsequent conversion step. During conversion, yellowcake is chemically processed to convert it into uranium hexafluoride (UF6), a stable and easily transportable form. UF6 is a crucial intermediate product used in the enrichment process to increase the concentration of the uranium-235 isotope, which is necessary for nuclear reactors.

Mining and processing uranium require careful consideration of safety and environmental factors. Engineers in the field of energy engineering play a vital role in ensuring that these operations are conducted in a responsible and sustainable manner. They must implement measures to mitigate the potential environmental impact, such as proper waste management and the prevention of radioactive contamination.

In conclusion, the mining and processing of uranium are critical steps in the production of nuclear power. Engineers in the field of energy engineering must possess a thorough understanding of these processes to ensure the safe and efficient extraction of this valuable resource. By adhering to rigorous safety and environmental standards, engineers can contribute to the sustainable development and utilization of nuclear power systems.

Conversion and Enrichment of Uranium

The conversion and enrichment of uranium is a crucial step in the production of nuclear fuel for power generation. This subchapter aims to provide engineers specializing in energy engineering with a comprehensive understanding of the processes involved in converting and enriching uranium.

The first section explores the conversion of naturally occurring uranium ore into a form suitable for enrichment. Engineers will learn about the various methods employed to extract and purify uranium, such as leaching, solvent extraction, and precipitation. The chapter will delve into the chemical and physical processes involved in each method, highlighting their advantages and limitations. Additionally, the importance of maintaining strict environmental and safety standards during the conversion process will be emphasized.

The second section focuses on uranium enrichment, which involves increasing the concentration of the fissile isotope, uranium-235, in natural uranium. Engineers will gain a deep understanding of the different enrichment technologies, including gaseous diffusion, gas centrifugation, and laser-based methods. Each technology's principles, efficiency, and technical challenges will be discussed in detail, enabling engineers to make informed decisions when selecting the most suitable enrichment method for a particular project.

Moreover, the subchapter will address the critical issue of nuclear proliferation and the role of enrichment in this context. It will highlight the international safeguards and regulations in place to prevent the misuse of enriched uranium for weapons purposes.

Engineers will gain insight into the challenges associated with maintaining a delicate balance between promoting peaceful uses of nuclear energy and ensuring non-proliferation.

Furthermore, the subchapter will explore the economic aspects of uranium conversion and enrichment. Engineers will be provided with an overview of the cost factors involved in these processes, such as energy consumption, capital investment, and operational expenses. The chapter will present cost-effective strategies and innovative technologies that can enhance the efficiency and sustainability of uranium conversion and enrichment.

Overall, this subchapter on conversion and enrichment of uranium will equip engineers specializing in energy engineering with the knowledge and skills necessary to contribute effectively to the design, operation, and optimization of nuclear power systems. By understanding the intricacies of these processes, engineers can play a vital role in ensuring the safe, reliable, and sustainable generation of nuclear power, contributing to the global energy transition and the mitigation of climate change.

Fuel Fabrication and Reactor Loading

In the realm of nuclear power systems, fuel fabrication and reactor loading are integral components of the energy engineering field. As engineers, it is crucial to understand the intricate processes involved in these stages to ensure the safe and efficient operation of nuclear reactors. This subchapter aims to provide a comprehensive overview of fuel fabrication and reactor loading, shedding light on their principles and applications.

Fuel fabrication, the first step in the nuclear fuel cycle, involves transforming raw materials into fuel assemblies suitable for nuclear reactors. Engineers involved in fuel fabrication must possess a deep understanding of materials science and manufacturing techniques. They work with enriched uranium or plutonium to create fuel rods or assemblies, employing precise measurements and rigorous quality control measures to ensure the fuel's integrity.

Once the fuel fabrication process is completed, the next crucial step is reactor loading. This process involves the careful insertion of fuel assemblies into the reactor core. Engineers must consider various factors, such as reactor design, power output requirements, and safety regulations, while loading the reactor. They meticulously plan and execute the loading sequence to optimize reactor performance and minimize the risk of accidents.

During reactor loading, engineers must also consider the concept of burnup, which refers to the amount of energy extracted from the fuel before it is replaced. The goal is to achieve an optimal burnup level while ensuring the fuel's structural integrity. This requires a thorough understanding of fuel behavior under different operating conditions and the ability to analyze data collected during reactor operation.

Furthermore, fuel fabrication and reactor loading are closely intertwined with safety considerations. Engineers must adhere to stringent safety protocols, including radiation protection measures and the prevention of criticality accidents. They also play a vital role in the design and implementation of safety systems, such as emergency shutdown mechanisms and containment structures, to mitigate the consequences of any potential incidents.

In conclusion, fuel fabrication and reactor loading form the foundation of nuclear power systems. Engineers specializing in energy engineering must possess a strong grasp of these processes to ensure the safe and efficient operation of nuclear reactors. By understanding the principles and applications of fuel fabrication and reactor loading, engineers can contribute to the advancement of nuclear power technology and meet the ever-increasing global energy demands.

Spent Fuel Management and Disposal

In the field of energy engineering, nuclear power has emerged as a significant source of electricity generation due to its efficiency, reliability, and low greenhouse gas emissions. However, the use of nuclear power also presents unique challenges, one of which is the management and disposal of spent fuel.

Spent fuel, also known as nuclear waste, refers to the fuel elements that have been used in a nuclear reactor and are no longer capable of sustaining a chain reaction. These fuel elements contain highly radioactive materials, which pose potential risks to human health and the environment if not managed properly.

The subchapter "Spent Fuel Management and Disposal" aims to provide engineers with a comprehensive understanding of the strategies and technologies involved in handling and disposing of nuclear waste. It covers various aspects, including the characteristics of spent fuel, current management practices, and future disposal options.

One of the key topics discussed in this subchapter is the storage of spent fuel. Engineers will learn about the different types of storage facilities, such as on-site storage pools and dry cask storage, and their advantages and limitations. The importance of maintaining the integrity and safety of these storage facilities is emphasized, considering the long lifespan of radioactive materials.

Moreover, the subchapter delves into the concept of reprocessing, which involves extracting useful materials from spent fuel for reuse. Engineers will gain insights into the different reprocessing techniques, their benefits, and associated challenges. This section also explores the international perspectives on reprocessing and the regulatory frameworks governing this practice.

Additionally, the subchapter addresses the long-term disposal of spent fuel. Engineers will be introduced to the concept of deep geological repositories, where nuclear waste can be safely stored for thousands of years. The technical and societal aspects of selecting suitable repository sites and ensuring long-term safety are explored in detail.

The subchapter concludes with an overview of emerging technologies and research efforts aimed at improving spent fuel management and disposal. Engineers will be encouraged to stay informed about advancements in this field and contribute to the development of innovative solutions.

Overall, "Spent Fuel Management and Disposal" offers engineers in the niche of energy engineering a valuable resource to understand the complex issues surrounding nuclear waste and equips them with the knowledge to design, implement, and improve strategies for the safe handling and disposal of spent fuel.

Chapter 5: Nuclear Reactor Operations

Startup and Shutdown Procedures

In the field of energy engineering, startup and shutdown procedures play a critical role in the safe and efficient operation of nuclear power systems. This subchapter aims to provide engineers with a comprehensive understanding of the procedures involved in starting up and shutting down a nuclear power plant.

Startup procedures are essential to ensure a smooth transition from an inactive state to full power generation. The startup process typically begins with pre-startup checks and inspections, ensuring that all systems, components, and safety features are functioning properly. These checks include verification of coolant flow, pressure, and temperature, as well as fuel integrity.

Once the pre-startup checks are complete, the reactor can be brought to a critical state. This is achieved by gradually increasing the power level while maintaining strict control over various parameters, such as coolant flow rate, reactor power, and neutron flux. The startup process requires precise coordination and monitoring to avoid any potential safety hazards, such as fuel cladding failure or excessive temperature rise.

Furthermore, engineers must be aware of the importance of maintaining proper coolant chemistry during startup. This involves careful control of water chemistry parameters, such as pH, conductivity, and dissolved oxygen levels. Proper coolant chemistry is crucial for preventing corrosion, minimizing radioactive material buildup, and ensuring the longevity of the reactor components.

On the other hand, shutdown procedures are equally critical in maintaining the safety and integrity of a nuclear power plant. The shutdown process is initiated when the reactor power needs to be reduced or halted completely. This can be due to routine maintenance, refueling, or in response to abnormal operating conditions.

During a shutdown, engineers must follow a carefully planned sequence of steps to bring the reactor to a safe and stable state. This involves gradually reducing the reactor power, adjusting control rods to maintain subcritical conditions, and monitoring various parameters to ensure the plant's integrity. Additionally, engineers must implement proper cooling, ventilation, and radiation monitoring measures during the shutdown process to safeguard personnel and the environment.

In conclusion, startup and shutdown procedures are crucial components of nuclear power system operations. Engineers specializing in energy engineering must possess a thorough understanding of these procedures to ensure safe operation, efficient power generation, and compliance with regulatory requirements. By following established guidelines and protocols, engineers can successfully navigate the complexities associated with startups and shutdowns, contributing to the sustainable and reliable operation of nuclear power plants.

Reactor Power Control and Load Following

In the field of nuclear power systems, reactor power control and load following are crucial concepts that engineers need to understand and implement effectively. This subchapter will delve into the principles and applications of reactor power control and load following, catering specifically to energy engineering professionals.

Reactor power control refers to the ability to adjust the power output of a nuclear reactor to meet the changing demands of the electrical grid. This control is achieved through various mechanisms, such as control rods, neutron poisons, and feedback systems. Engineers play a pivotal role in designing, developing, and fine-tuning these control mechanisms to ensure safe and efficient operation of nuclear power plants.

Load following, on the other hand, involves the reactor's ability to adjust its power output in response to fluctuations in electricity demand. This capability is of utmost importance in modern power systems that heavily rely on renewable energy sources, which are inherently intermittent in nature. Engineers must devise sophisticated control strategies that allow nuclear reactors to smoothly transition between different power levels while maintaining stability and safety.

The subchapter will first provide an overview of the basic principles behind reactor power control and load following. It will delve into the physics of nuclear reactors, explaining how control rods and other mechanisms are used to regulate the fission process and adjust power output. The importance of feedback systems and reactor dynamics will also be highlighted, emphasizing the need for precise control and monitoring.

Next, the subchapter will explore advanced control strategies and technologies employed in modern nuclear power systems. Topics such as digital control systems, predictive control algorithms, and adaptive control methods will be discussed in-depth, showcasing the cutting-edge advancements in reactor power control and load following.

Furthermore, the subchapter will delve into the safety considerations associated with reactor power control and load following. It will address issues like reactivity accidents, overheating, and the impact of load following on fuel integrity. Engineers will gain insights into the safety measures and protocols that must be implemented to mitigate these risks effectively.

In conclusion, this subchapter on reactor power control and load following will equip energy engineering professionals with the knowledge and tools necessary to handle the complex dynamics of nuclear power systems. With a focus on safety, efficiency, and adaptability, engineers will be able to contribute to the sustainable and reliable operation of nuclear power plants in the rapidly evolving energy landscape.

Reactor Safety and Emergency Procedures

In the field of energy engineering, nuclear power systems are widely recognized for their ability to generate vast amounts of electricity with minimal greenhouse gas emissions. However, the potential risks associated with nuclear power plants necessitate a comprehensive understanding of reactor safety and emergency procedures. This subchapter aims to provide engineers with the necessary knowledge and skills to ensure the safe and reliable operation of nuclear power systems.

The subchapter begins by exploring the fundamentals of reactor safety, emphasizing the various defense mechanisms in place to prevent accidents and mitigate their consequences. Topics such as inherent safety features, redundancy in safety systems, and the defense-in-depth approach are discussed in detail. Through a systematic analysis of reactor design and safety measures, engineers can develop a thorough understanding of how nuclear power systems operate under normal conditions.

Furthermore, this subchapter delves into emergency procedures, highlighting the importance of preparedness in handling unforeseen events. Engineers are provided with insight into the various types of emergencies that can occur in a nuclear power plant, including loss of coolant accidents, power supply failures, and extreme weather events. Detailed procedures for each scenario are outlined, ensuring that engineers are equipped to respond swiftly and effectively.

The content also covers the role of emergency response organizations and their collaboration with plant operators during critical situations.

It emphasizes the significance of communication, coordination, and decision-making in ensuring the safety of personnel and the surrounding environment. Engineers are encouraged to familiarize themselves with emergency response protocols and participate in regular drills and exercises to enhance preparedness.

In addition to emergency response, this subchapter addresses post-accident management and recovery. Engineers are introduced to techniques for assessing and mitigating the consequences of accidents, such as containment strategies, environmental monitoring, and public information dissemination. By understanding the long-term implications of accidents and implementing appropriate measures, engineers can contribute to the restoration of normal operations and public confidence in nuclear power systems.

Overall, this subchapter on reactor safety and emergency procedures offers engineers in the field of energy engineering a comprehensive guide to ensuring the safe and reliable operation of nuclear power systems. By equipping themselves with the knowledge and skills outlined in this subchapter, engineers can effectively address potential risks, respond to emergencies, and contribute to the continued development and success of nuclear power as a clean and sustainable energy source.

Maintenance and Refueling Operations

In the realm of nuclear power systems, maintenance and refueling operations are crucial aspects that engineers must consider to ensure the safe and reliable operation of these complex energy generation facilities. This subchapter delves into the various aspects of maintenance and refueling operations in nuclear power systems, offering valuable insights and guidelines for energy engineers.

Maintenance is an integral part of nuclear power system operations, as it helps identify and mitigate potential issues before they escalate into major problems. Regular inspections, testing, and maintenance activities are carried out to ensure the integrity of the components and systems, thereby guaranteeing the safe and efficient functioning of the plant. Engineers in the field of energy engineering need to possess a comprehensive understanding of the maintenance procedures, including preventive, corrective, and predictive maintenance strategies.

Preventive maintenance involves scheduled inspections, cleaning, lubrication, and replacement of equipment or components at predetermined intervals. By adhering to a proactive maintenance approach, engineers can minimize the likelihood of unexpected failures, optimize the system's performance, and extend the lifespan of critical components. Corrective maintenance, on the other hand, focuses on addressing malfunctions or failures promptly to prevent further damage or downtime. Predictive maintenance techniques, such as condition monitoring and performance analysis, aid in identifying potential issues based on system behavior, enabling engineers to take preventive measures.

Refueling operations in nuclear power systems are another critical aspect that engineers must master. Typically, nuclear reactors require periodic refueling to replace spent fuel rods and ensure continuous power generation. Energy engineers play a vital role in planning and executing refueling operations, ensuring safety, efficiency, and minimal downtime. The process involves careful handling, transportation, and storage of radioactive materials, necessitating strict adherence to regulatory guidelines and safety protocols.

This subchapter also explores emerging trends and advancements in maintenance and refueling operations, such as robotics and automation. The integration of robotic systems allows for safer and more efficient inspections, repairs, and refueling processes, reducing human exposure to radiation and enhancing overall operational performance.

Overall, this subchapter on maintenance and refueling operations provides energy engineers with valuable insights into the best practices, techniques, and technologies involved in ensuring the smooth functioning of nuclear power systems. By mastering these concepts, engineers can contribute to the sustainable and reliable generation of nuclear energy, meeting the growing demands of the energy engineering niche.

Chapter 6: Thermal Hydraulics and Heat Transfer in Nuclear Systems

Fluid Mechanics in Nuclear Systems

Fluid mechanics plays a crucial role in the efficient operation and safety of nuclear power systems. Understanding the principles and applications of fluid mechanics is essential for engineers working in the field of energy engineering, specifically those involved in nuclear power systems. This subchapter aims to provide a comprehensive overview of the key concepts and their relevance to nuclear systems.

In nuclear power systems, fluid mechanics is primarily concerned with the behavior of liquids and gases as they flow through various components and systems. It encompasses the study of fluid properties, fluid dynamics, and the principles of heat transfer. Engineers in the field of energy engineering need to grasp these fundamental concepts to design, operate, and maintain nuclear power plants.

One of the key areas where fluid mechanics is applied in nuclear systems is in the design and analysis of coolant systems. Coolant, typically in the form of water, is used to remove heat generated in the nuclear reactor. Understanding fluid dynamics helps engineers optimize the flow rate, pressure, and temperature distribution within the coolant system, ensuring efficient heat transfer and preventing any hotspots that could lead to system failure.

Another crucial application of fluid mechanics in nuclear systems is in the design of safety systems. For instance, emergency core cooling systems are designed to prevent core meltdown in the event of a loss of coolant accident. Fluid mechanics principles are used to determine the flow rates, pressures, and locations of emergency coolant injections to ensure rapid and effective cooling of the reactor core.

Additionally, fluid mechanics is essential in the design and analysis of steam generators and condensers. These components play a critical role in converting thermal energy from the reactor into electrical energy. Understanding fluid properties and heat transfer principles enables engineers to optimize the design and performance of these components, enhancing overall plant efficiency.

In conclusion, fluid mechanics is a vital discipline within the field of energy engineering, particularly in the context of nuclear power systems. Its principles and applications are indispensable for engineers involved in the design, operation, and maintenance of nuclear power plants. By understanding fluid dynamics, heat transfer, and fluid properties, engineers can ensure efficient and safe operation of nuclear systems while maximizing energy output.

Heat Transfer Mechanisms

In the field of energy engineering, understanding heat transfer mechanisms is crucial for designing and optimizing nuclear power systems. Heat transfer plays a fundamental role in the efficient operation and safety of these systems, and engineers must have a comprehensive knowledge of the different mechanisms involved.

The primary heat transfer mechanisms encountered in nuclear power systems are conduction, convection, and radiation. Each mechanism has its unique characteristics and influences the overall heat transfer process differently.

Conduction is the transfer of heat through a solid material or between two objects in direct contact. In nuclear power systems, conduction occurs within components such as fuel rods, reactor walls, and heat exchangers. Engineers must carefully consider the thermal conductivity and thickness of these materials to ensure efficient heat transfer and prevent overheating.

Convection involves the transfer of heat through a fluid, either in the form of forced convection or natural convection. Forced convection occurs when a fluid is forced to flow over a surface, such as in heat exchangers or coolant systems. Natural convection, on the other hand, relies on buoyancy forces to circulate the fluid. Engineers must account for factors such as fluid properties, flow rates, and pressure drops to optimize convection heat transfer in nuclear power systems.

Radiation is the transfer of heat through electromagnetic waves, without the need for a medium. In nuclear power systems, radiation plays a significant role in heat transfer between the reactor core and the surrounding environment. Engineers must consider factors such as surface emissivity, temperature differentials, and shielding materials to control and manage radiation heat transfer effectively.

The choice of heat transfer mechanism in nuclear power systems depends on various factors such as system design, operating conditions, and safety requirements. Engineers must analyze and evaluate these factors to select the most suitable mechanism or combination of mechanisms for a given application.

Additionally, engineers must also understand heat transfer enhancement techniques to improve the efficiency of nuclear power systems. These techniques include the use of fins, turbulators, heat pipes, and advanced cooling fluids. By employing these enhancements, engineers can optimize heat transfer rates and improve overall system performance.

In conclusion, heat transfer mechanisms are essential considerations in the design and operation of nuclear power systems. Engineers specializing in energy engineering must have a comprehensive understanding of conduction, convection, and radiation heat transfer mechanisms to ensure the efficient and safe operation of these systems. Moreover, knowledge of heat transfer enhancement techniques is crucial for maximizing the performance and reliability of nuclear power systems in the field of energy engineering.

Heat Exchangers and Steam Generator Design

In the realm of energy engineering, heat exchangers and steam generator design play a pivotal role in the efficient and safe operation of nuclear power systems. This subchapter aims to provide engineers with a comprehensive understanding of the principles and applications of these crucial components.

Heat exchangers are vital devices that facilitate the transfer of thermal energy between two or more fluids at different temperatures. In the context of nuclear power systems, heat exchangers are primarily used to extract heat from the reactor coolant and transfer it to the secondary coolant. This process helps in generating steam, which drives the turbines to produce electricity.

The design of heat exchangers is a complex task that necessitates careful consideration of numerous factors. Engineers need to analyze the heat transfer characteristics of the fluids involved, select appropriate materials, and determine the optimal configuration and size of the exchanger. Additionally, considerations such as pressure drop, fouling, and maintenance requirements must be taken into account to ensure the long-term performance and reliability of the heat exchanger.

Steam generators, on the other hand, are integral components in nuclear power plants that convert water into steam by utilizing the heat transferred from the primary coolant. Efficient steam generator design is crucial to maximize the energy conversion process while maintaining safety and reliability. Engineers need to consider factors such as tube material, geometry, and arrangement to ensure efficient heat transfer and minimize the risk of corrosion or tube failure.

In this subchapter, engineers will delve into the various types of heat exchangers commonly used in nuclear power systems, including shell-and-tube, plate, and finned tube heat exchangers. The advantages, disadvantages, and specific design considerations for each type will be explored, along with practical examples and case studies.

Furthermore, engineers will gain insight into the design principles and operational considerations of steam generators. Key topics covered will include flow distribution, steam quality control, and methods for mitigating potential issues such as vibration, flow-induced vibration, and tube fouling.

By providing engineers with a comprehensive understanding of heat exchangers and steam generator design, this subchapter aims to equip them with the knowledge and tools necessary to optimize the performance, efficiency, and safety of nuclear power systems.

Containment Systems and Pressure Control

In the field of energy engineering, the safety and reliability of nuclear power systems are of paramount importance. One critical aspect of ensuring this safety is the implementation of effective containment systems and pressure control measures. These systems are designed to prevent the release of radioactive materials in the event of an accident or abnormal operating conditions.

Containment systems serve as a physical barrier to confine radioactive materials within the reactor and prevent their escape into the environment. They are typically constructed using reinforced concrete and steel, with multiple layers of protection. The primary containment structure surrounds the reactor vessel, while the secondary containment provides an additional layer of defense. These structures are built to withstand extreme external events such as earthquakes, tornadoes, and even aircraft crashes.

The pressure within the containment is a crucial parameter that needs to be carefully controlled. Excessive pressure can lead to structural failure, while too low pressure can result in the ingress of air and other contaminants. To maintain the desired pressure conditions, various pressure control systems are employed.

One of the key components of pressure control systems is the emergency core cooling system (ECCS). ECCS provides a means to cool the reactor core in case of a loss of coolant accident, preventing overheating and potential meltdown. This system utilizes pumps, heat exchangers, and storage tanks to circulate and cool the coolant, ensuring optimal pressure levels are maintained.

Another important pressure control mechanism is the containment venting system. This system allows for the controlled release of pressure and steam from the containment during abnormal conditions. By venting the excess pressure, the risk of containment failure is minimized. However, this process needs to be carefully managed to prevent the release of radioactive materials into the environment.

The design and implementation of effective containment systems and pressure control measures require thorough analysis, extensive testing, and adherence to strict regulatory standards. Engineers play a critical role in ensuring the safe and efficient operation of nuclear power plants. Their expertise in energy engineering is essential for developing robust containment systems and pressure control strategies that can withstand both anticipated and unforeseen events.

By continuously improving and innovating in this area, engineers contribute to the overall advancement of nuclear power systems, making them more reliable, sustainable, and environmentally friendly.

Chapter 7: Radiation Protection and Shielding

Radiation Sources and Types

In the field of energy engineering, understanding radiation sources and types is crucial for engineers working with nuclear power systems. This subchapter aims to provide a comprehensive overview of the various sources of radiation and the types of radiation encountered in this field.

Radiation, in the context of nuclear power systems, refers to the emission of energy in the form of electromagnetic waves or subatomic particles. It is a fundamental aspect of nuclear processes and plays a significant role in power generation and other applications.

There are two primary sources of radiation in nuclear power systems: natural and artificial sources. Natural sources include cosmic rays from space, radioactive materials present in the Earth's crust, and solar radiation. Artificial sources, on the other hand, are produced through human activities such as nuclear power generation, medical procedures, industrial applications, and research facilities.

Radiation can be classified into two main types: ionizing and non-ionizing radiation. Ionizing radiation possesses sufficient energy to remove tightly bound electrons from atoms, leading to ionization. This type of radiation includes alpha particles, beta particles, gamma rays, and X-rays. Engineers need to be aware of the properties and characteristics of ionizing radiation, as it poses potential health risks and requires appropriate shielding and safety measures.

Non-ionizing radiation, on the other hand, lacks the energy to ionize atoms but can still cause biological effects. This type includes radio waves, microwaves, and visible light. While non-ionizing radiation is generally considered less hazardous, engineers should still understand its properties and potential interactions with materials and systems.

It is crucial for engineers to have a deep understanding of radiation sources and types to ensure the safe and efficient operation of nuclear power systems. This knowledge enables them to design appropriate shielding, implement effective safety protocols, and accurately assess potential risks associated with radiation exposure.

In conclusion, the subchapter "Radiation Sources and Types" provides engineers specializing in energy engineering with essential knowledge about the various sources and types of radiation encountered in nuclear power systems. By understanding the properties, characteristics, and potential hazards of radiation, engineers can make informed decisions and develop effective strategies to ensure the safety and reliability of these systems.

In the field of energy engineering, understanding radiation sources and types is crucial for engineers working with nuclear power systems. This subchapter delves into the fundamental concepts of radiation, its sources, and various types encountered in nuclear power applications.

Radiation, in the context of nuclear power systems, refers to the emission of energy in the form of particles or electromagnetic waves. These emissions occur due to the unstable nature of atomic nuclei, leading to their transformation into more stable configurations. Engineers working with nuclear power systems must be well-versed in radiation sources and types to ensure the safe and efficient operation of these systems.

The primary source of radiation encountered in nuclear power systems is radioactive materials. These materials, such as uranium and plutonium, have unstable atomic nuclei that undergo radioactive decay, releasing energy in the form of radiation. Radioactive decay can occur through several processes, including alpha, beta, and gamma decay.

Alpha decay involves the emission of alpha particles, which are composed of two protons and two neutrons. These particles have a positive charge and relatively low penetrating power, making them easily stopped by a few centimeters of air or a sheet of paper. Beta decay, on the other hand, involves the emission of beta particles, which can be either electrons (beta-minus) or positrons (beta-plus). Beta particles have higher penetrating power than alpha particles and can be stopped by a few millimeters of aluminum or plastic.

Gamma decay, unlike alpha and beta decay, does not involve the emission of particles. Instead, it releases high-energy electromagnetic waves called gamma rays. Gamma rays have the highest penetrating power among the three types of radiation and require several centimeters of lead or concrete to be effectively shielded.

Understanding the characteristics and behavior of different types of radiation is essential for engineers when designing shielding systems and safety measures to protect workers and the environment from potential radiation hazards. By comprehending the sources and types of radiation encountered in nuclear power systems, engineers can effectively manage and mitigate risks associated with radiation exposure.

In conclusion, this subchapter provides engineers specializing in energy engineering with a comprehensive overview of radiation sources and types in nuclear power systems. By understanding the fundamentals of radiation and its various forms, engineers can make informed decisions when designing and operating nuclear power plants, ensuring the safety and efficiency of these crucial energy systems.

Biological Effects of Radiation Exposure

Radiation exposure is a critical concern when it comes to nuclear power systems, and engineers in the field of energy engineering must have a thorough understanding of its biological effects. This subchapter aims to provide engineers with an overview of the potential health risks associated with radiation exposure and the measures taken to mitigate them.

Radiation can have detrimental effects on human health due to its ability to ionize atoms and molecules in the body. When ionizing radiation interacts with living tissue, it can cause DNA damage, leading to various health conditions, including cancer and genetic disorders. It is crucial for engineers working on nuclear power systems to be aware of these potential risks and take appropriate precautions to minimize radiation exposure.

Firstly, engineers must understand the concept of dose measurement in radiation protection. The subchapter will delve into units such as the gray (Gy) and sievert (Sv), which are used to quantify the amount of radiation absorbed by the body. Engineers will also learn about the different types of ionizing radiation, including alpha particles, beta particles, gamma rays, and neutrons, and their varying levels of penetration and potential harm.

The subchapter will then explore the biological effects of radiation exposure at different dose levels. It will cover acute radiation syndrome (ARS), which occurs when the body receives a high dose of radiation over a short period. Engineers will learn about the symptoms and immediate effects of ARS, as well as the long-term consequences for survivors.

Furthermore, the subchapter will discuss the concept of stochastic effects, which are associated with long-term, low-dose radiation exposure. These effects, such as cancer and genetic mutations, occur randomly and have a probability of occurrence that increases with increasing radiation dose. Engineers will gain an understanding of the dose thresholds for various stochastic effects and the importance of dose limits in ensuring radiation safety.

To conclude, the subchapter will emphasize the importance of implementing strict radiation safety practices in nuclear power systems. Engineers will learn about the design principles and engineering controls that are in place to minimize radiation exposure for workers and the general public. This knowledge will enable engineers to make informed decisions and contribute to the safe and sustainable operation of nuclear power systems within the field of energy engineering.

In summary, this subchapter on the biological effects of radiation exposure provides engineers in the niche of energy engineering with essential knowledge about the potential health risks associated with radiation exposure. By understanding the dose measurement, types of ionizing radiation, acute and stochastic effects, and radiation safety practices, engineers will be equipped to design and operate nuclear power systems with a strong emphasis on minimizing radiation exposure and ensuring the safety of both workers and the general public.

In the field of energy engineering, where the focus is on harnessing and utilizing power sources efficiently, it is crucial to understand the potential biological effects of radiation exposure. As engineers working with nuclear power systems, it is our responsibility to ensure the safety of both the environment and human health.

Radiation, in the context of nuclear power, refers to the emission of energy as electromagnetic waves or as moving subatomic particles. While radiation has numerous beneficial applications, including electricity production and medical imaging, it also poses potential risks to living organisms.

One of the primary concerns when dealing with radiation exposure is the damage it can cause to human cells. Ionizing radiation, such as gamma rays and X-rays, has enough energy to remove tightly bound electrons from atoms, leading to the formation of charged particles. These charged particles, known as ions, can disrupt chemical bonds within cells and DNA, potentially causing genetic mutations or cell death.

The severity of biological effects depends on several factors, including the type of radiation, the dose received, and the duration of exposure. Acute exposure to high doses of radiation can lead to immediate health effects, such as radiation sickness, burns, and even death. On the other hand, chronic exposure to lower doses of radiation may increase the risk of cancer and other long-term health issues.

Engineers working with nuclear power systems must adhere to strict safety protocols to minimize radiation exposure. These protocols involve the use of shielding materials, proper containment of radioactive substances, and regular monitoring of radiation levels. Additionally, engineers play a crucial role in designing and maintaining safety systems, such as emergency shutdown mechanisms and radiation detection devices.

Understanding the biological effects of radiation exposure is essential not only for the safety of those working with nuclear power systems but also for the general public. Engineers in the field of energy engineering have a responsibility to educate society about the potential risks associated with radiation and the measures taken to mitigate them.

By continuously advancing our knowledge of radiation biology and safety practices, engineers can ensure the sustainable and responsible use of nuclear power systems, while minimizing the potential harm to both humans and the environment.

Shielding Materials and Design

In the field of nuclear power systems, one of the most critical aspects is the effective shielding of radiation. Shielding materials and design play a vital role in ensuring the safety and efficient operation of nuclear power plants. This subchapter aims to provide engineers in the niche of energy engineering with an in-depth understanding of shielding materials and their design principles.

Shielding materials are essential in reducing the harmful effects of radiation by absorbing or scattering the radiation particles. The choice of suitable shielding materials depends on various factors, including the type and energy of radiation, the required shielding thickness, and cost-effectiveness. Commonly used shielding materials include concrete, lead, steel, and water. Each material has its specific advantages and limitations, making it crucial to select the most appropriate material for each application.

Concrete is widely used in nuclear power systems due to its excellent shielding properties, cost-effectiveness, and ease of construction. Its high density and composition make it effective in attenuating gamma rays and neutrons. Additionally, concrete can be easily molded into various shapes, allowing for flexible shielding designs.

Lead is another commonly used shielding material due to its high atomic number, which enables effective attenuation of gamma rays. However, lead is not suitable for neutron shielding, as it tends to scatter neutrons rather than absorbing them. Steel, on the other hand, is an excellent choice for neutron shielding due to its high absorption cross-section for neutrons. It is often used in combination with lead or concrete to provide comprehensive radiation shielding.

Water is frequently employed as a shielding material in nuclear power systems, particularly for the cooling and moderation of neutrons. Its high hydrogen content makes it effective in slowing down fast neutrons. Water is also used as a radiation shield in spent fuel pools, where it not only absorbs radiation but also provides a cooling medium.

Designing effective shielding requires careful consideration of factors such as radiation source location, required shielding thickness, and the surrounding environment. The shielding design should be optimized to minimize radiation exposure to personnel while ensuring the structural integrity of the system.

In conclusion, shielding materials and design are crucial aspects of nuclear power systems. Engineers in the field of energy engineering must have a comprehensive understanding of various shielding materials and their characteristics. By selecting the appropriate materials and designing effective shielding structures, engineers can ensure the safety and efficiency of nuclear power plants, thereby contributing to the sustainable development of energy resources.

In the realm of nuclear power systems, the safety of personnel and the surrounding environment is of paramount importance. Shielding materials and design play a crucial role in ensuring the effective containment of radiation and preventing its harmful effects. This subchapter delves into the intricate world of shielding materials and their design principles, providing engineers in the field of energy engineering with essential knowledge and tools to address the challenges of nuclear power systems.

The first section of this subchapter explores the characteristics and properties of shielding materials. Engineers must understand the fundamental principles underlying shielding materials to make informed decisions in their design choices. Topics covered include the attenuation of radiation, the concept of linear attenuation coefficient, and the different types of shielding materials available. Special emphasis is placed on materials such as lead, concrete, and steel, which are commonly used in shielding applications.

The subsequent section delves into the design considerations for efficient shielding. Engineers need to optimize the shielding design to achieve the desired levels of radiation attenuation while considering factors such as space limitations, cost-effectiveness, and ease of maintenance. This section introduces the concept of shielding thickness calculations, taking into account the radiation source strength, the desired dose reduction, and the specific properties of selected shielding materials.

To provide engineers with practical insights, the subchapter also presents case studies of successful shielding designs in nuclear power systems. These case studies highlight the importance of a multidisciplinary approach, involving experts in nuclear physics, material science, and structural engineering. Engineers will gain valuable insights into the challenges faced during the design process and the innovative solutions employed to achieve effective radiation shielding.

Furthermore, the subchapter discusses the advances in shielding materials, including novel composites and alloys that offer enhanced radiation attenuation properties. It also explores emerging technologies such as active shielding, which employs real-time monitoring and feedback systems to dynamically adjust shielding effectiveness.

In conclusion, "Shielding Materials and Design" is an essential subchapter within the book "Nuclear Power Systems: Principles and Applications for Engineers." It equips energy engineering professionals with the knowledge and tools needed to design efficient and effective shielding solutions in the context of nuclear power systems. By understanding the characteristics of shielding materials, optimizing design considerations, and exploring the latest advancements in the field, engineers can ensure the safety and integrity of nuclear power systems for a sustainable and secure future.

Occupational Radiation Protection

Radiation protection is of paramount importance in the field of nuclear power systems, especially when it comes to ensuring the safety of workers and engineers involved in the energy engineering industry. This subchapter aims to provide a comprehensive overview of occupational radiation protection, including the principles, regulations, and best practices that engineers must adhere to in order to safeguard themselves and others from potential radiation hazards.

One of the fundamental principles of occupational radiation protection is the concept of ALARA, which stands for "As Low As Reasonably Achievable." This principle emphasizes the need to minimize radiation exposure by implementing stringent safety measures and adopting innovative technologies. Engineers play a crucial role in designing and implementing these safety measures, such as shielding materials, radiation monitoring systems, and proper waste management protocols.

Regulatory bodies, such as the International Atomic Energy Agency (IAEA) and national nuclear regulatory agencies, establish guidelines and regulations to ensure the protection of workers in the nuclear industry. Engineers must stay informed about these regulations and implement them in their daily work. This includes conducting regular risk assessments, setting dose limits, and providing appropriate training to workers on radiation safety practices.

Personal protective equipment (PPE) is another critical aspect of occupational radiation protection. Engineers working in nuclear power systems must wear specialized PPE, including lead aprons, gloves, and goggles, to shield themselves from ionizing radiation. Regular inspection and maintenance of PPE is necessary to ensure its effectiveness and prevent any potential risks.

Furthermore, engineers must be well-versed in the use of radiation monitoring devices. These devices, such as dosimeters and Geiger-Muller counters, help measure radiation levels and ensure that workers are not exposed to excessive radiation. Regular monitoring and analysis of radiation data are essential for identifying potential sources of radiation and taking appropriate corrective actions.

In addition to these technical aspects, engineers should also prioritize communication and collaboration in occupational radiation protection. Regular safety briefings, training sessions, and open dialogue among workers can help create a safety culture and promote awareness of radiation hazards. This collaborative approach ensures that all workers, including engineers, are actively involved in maintaining a safe working environment.

In conclusion, occupational radiation protection is a critical component of the nuclear power systems industry, and engineers in the field of energy engineering must be well-versed in its principles and practices. By adhering to regulatory guidelines, implementing ALARA principles, utilizing PPE, monitoring radiation levels, and promoting a culture of safety, engineers can effectively protect themselves and their colleagues from the potential hazards associated with radiation exposure.

Radiation is an integral part of nuclear power systems, and as engineers in the field of energy engineering, it is crucial to have a comprehensive understanding of occupational radiation protection. This subchapter aims to provide engineers with the necessary knowledge and guidelines to ensure the safety of personnel working in nuclear power plants and other related facilities.

First and foremost, it is essential to comprehend the potential risks associated with radiation exposure. Ionizing radiation can have harmful effects on human health, including damage to DNA and the risk of developing cancer. Thus, engineers must be well-versed in the principles of radiation protection to minimize these risks effectively.

One of the fundamental principles is the ALARA (As Low As Reasonably Achievable) concept. This concept emphasizes the importance of minimizing radiation exposure by implementing appropriate control measures. Engineers should be familiar with the various techniques and technologies available to reduce occupational radiation exposure, such as shielding, time, distance, and the use of personal protective equipment (PPE).

There are also regulatory bodies and international standards that govern occupational radiation protection. Understanding these regulations is vital for engineers to ensure compliance and maintain a safe working environment. The subchapter will delve into the key regulatory organizations and standards, providing engineers with the necessary information to navigate this complex landscape effectively.

Additionally, the subchapter will cover topics such as radiation monitoring and dosimetry. Engineers need to be familiar with the different types of monitoring devices, their applications, and how to interpret the data obtained. Dosimetry, the measurement of radiation dose, is crucial for assessing the potential risks and ensuring that exposure limits are not exceeded.

Furthermore, the subchapter will address emergency preparedness and response in the context of occupational radiation protection. Nuclear power plants are equipped with emergency plans and procedures to mitigate the impact of potential accidents. Engineers must be well-versed in these plans and understand their roles and responsibilities in such situations.

In conclusion, occupational radiation protection is a critical aspect of working in the field of energy engineering, particularly in the nuclear power sector. This subchapter aims to equip engineers with the necessary knowledge and guidelines to ensure the safety of personnel. By understanding the principles, regulations, and best practices of radiation protection, engineers can effectively minimize the risks associated with radiation exposure and maintain a safe working environment.

Chapter 8: Nuclear Power Plant Economics and Environmental Impact

Cost Analysis of Nuclear Power Systems

In the field of energy engineering, it is crucial to analyze the cost implications of different power systems. This subchapter aims to provide engineers with a comprehensive understanding of the cost analysis of nuclear power systems. As nuclear power continues to play a significant role in the global energy sector, engineers need to evaluate the economic feasibility and sustainability of such systems.

Nuclear power systems offer several advantages, including high energy density, low greenhouse gas emissions, and a consistent power output. However, the initial capital investment required for constructing nuclear power plants is substantial. Engineers must consider factors such as the cost of land, equipment, material, and labor when estimating the overall cost of a nuclear power system.

One key aspect of cost analysis is the levelized cost of electricity (LCOE). Engineers must calculate the LCOE to determine the average cost of electricity generated by a nuclear power system over its lifetime. This calculation involves considering the initial capital costs, operating and maintenance expenses, fuel costs, and the duration of the plant's operation. By comparing the LCOE of nuclear power systems with other energy sources, engineers can assess their economic competitiveness.

Furthermore, engineers need to evaluate the impact of regulatory requirements and safety measures on the cost of nuclear power systems. Stringent safety standards are essential to prevent accidents, but they can significantly increase the overall cost. Thus, engineers must strike a balance between safety and cost-effectiveness while designing and implementing nuclear power systems.

Additionally, decommissioning costs must be factored into the analysis. Nuclear power plants have a limited operational lifespan, and their decommissioning involves dismantling and decontaminating the facility. Engineers must estimate these costs accurately to ensure the financial sustainability of nuclear power systems.

Lastly, engineers should consider the potential for technological advancements and economies of scale in nuclear power systems. Advancements in reactor designs, fuel efficiency, and waste management techniques can reduce costs and enhance the economic viability of these systems in the long run.

In conclusion, a comprehensive cost analysis is vital for engineers specializing in energy engineering to evaluate the feasibility and sustainability of nuclear power systems. By considering factors such as the levelized cost of electricity, regulatory requirements, safety measures, decommissioning costs, and technological advancements, engineers can make informed decisions regarding the economic viability of nuclear power systems. This subchapter aims to equip engineers with the knowledge and tools necessary to conduct a thorough cost analysis and contribute to the development of efficient and cost-effective nuclear power systems.

In the field of energy engineering, it is crucial to understand the cost implications of different power systems in order to make informed decisions. One such system that has gained significant attention is nuclear power. This subchapter aims to provide engineers with a comprehensive analysis of the costs associated with nuclear power systems.

When evaluating the cost of nuclear power systems, it is important to consider both the initial investment and the long-term operational costs. The initial investment includes the construction of the power plant, the procurement of nuclear fuel, and the installation of safety measures. Nuclear power plants require specialized infrastructure and safety features, which contribute significantly to the upfront costs. However, it is important to note that nuclear power plants have a longer lifespan compared to other power systems, resulting in a higher return on investment over time.

In terms of operational costs, nuclear power systems have several advantages. Nuclear fuel is relatively inexpensive compared to other sources such as fossil fuels. Additionally, nuclear power plants require a smaller workforce, resulting in lower labor costs. However, it is essential to consider the costs associated with waste management and decommissioning. Proper disposal and storage of nuclear waste require meticulous planning and investment, adding to the overall costs of nuclear power systems.

Furthermore, it is important to examine the economic viability of nuclear power in comparison to other energy sources. While nuclear power systems have higher initial costs, they offer a stable and reliable source of energy, reducing the dependency on fluctuating fuel prices. Additionally, nuclear power is considered a low-carbon option, aligning with the global push for sustainable energy solutions.

To make an accurate cost analysis, engineers must also consider external factors such as government policies and regulations. Governments play a significant role in providing financial incentives and subsidies to promote the development of nuclear power systems. These incentives can significantly impact the overall cost and profitability of nuclear power projects.

In conclusion, a comprehensive cost analysis is crucial when evaluating the feasibility and viability of nuclear power systems. Engineers in the field of energy engineering must consider both the initial investment and long-term operational costs, including waste management and decommissioning. Despite higher upfront costs, nuclear power systems offer stable and reliable energy generation with lower fuel expenses. Government policies and regulations should also be taken into account to obtain a realistic assessment. By understanding the cost implications, engineers can make informed decisions in developing and implementing nuclear power systems for a sustainable future.

Environmental Impacts of Nuclear Power

Nuclear power has long been hailed as a viable solution to the world's energy needs, offering substantial benefits such as reduced greenhouse gas emissions and a reliable source of electricity. However, it is crucial for engineers and professionals in the field of energy engineering to have a comprehensive understanding of the environmental impacts associated with nuclear power systems. This subchapter aims to shed light on these impacts, providing engineers with the knowledge necessary to make informed decisions and mitigate potential negative consequences.

One of the most significant environmental impacts of nuclear power is the generation of radioactive waste. Nuclear power plants produce various types of waste, including spent fuel rods and other radioactive materials, which require careful handling and disposal. Engineers play a crucial role in designing safe and efficient waste management systems to prevent any leakage or contamination of the environment. These systems often involve the use of specialized containers and storage facilities, as well as long-term plans for permanent waste repositories.

Another environmental concern is the potential for accidents and their consequences. While modern nuclear power plants incorporate multiple safety features and protocols, the risk of accidents can never be completely eliminated. Engineers must continuously work towards improving safety measures and emergency response plans to minimize the impacts of any potential incidents. This includes designing robust containment structures and implementing effective monitoring and control systems.

Beyond accidents, the mining and processing of uranium, the fuel used in nuclear power plants, can have adverse environmental effects. The extraction process can lead to habitat destruction, water pollution, and the release of harmful chemicals. Engineers need to explore alternative methods for uranium extraction that minimize environmental damage and promote sustainability.

Additionally, the thermal pollution of water bodies is another consideration. Nuclear power plants require large amounts of water for cooling purposes, and the heated water is often released back into rivers or lakes, impacting aquatic ecosystems. Engineers can mitigate these effects by implementing advanced cooling technologies, such as closed-loop cooling systems and cooling towers.

Lastly, the decommissioning of nuclear power plants at the end of their operational life poses its own set of environmental challenges. Engineers must develop strategies for the safe dismantling and disposal of radioactive materials, as well as the remediation of contaminated sites.

In conclusion, while nuclear power offers significant advantages in terms of reduced greenhouse gas emissions and reliable electricity generation, it also comes with its share of environmental impacts. Engineers specializing in energy engineering have a crucial role to play in understanding and mitigating these impacts. By developing innovative solutions for waste management, improving safety measures, exploring sustainable uranium extraction methods, implementing advanced cooling technologies, and planning for decommissioning, engineers can ensure that nuclear power systems are developed and operated in an environmentally responsible manner.

Nuclear power is a complex and controversial topic that has significant environmental implications. Engineers in the field of energy engineering need to have a comprehensive understanding of the environmental impacts associated with nuclear power in order to make informed decisions and develop sustainable solutions. This subchapter aims to provide engineers with an overview of the environmental impacts of nuclear power, addressing both the positive and negative aspects.

One of the key advantages of nuclear power is its low greenhouse gas emissions. Nuclear power plants generate electricity without combustion, which means they do not release carbon dioxide or other greenhouse gases into the atmosphere. This makes nuclear power a valuable tool in the fight against climate change, as it can help reduce global carbon emissions. Additionally, nuclear power plants have a high capacity factor, meaning they can produce electricity continuously, ensuring a stable and reliable power supply.

However, nuclear power is not without its drawbacks. The most significant environmental concern associated with nuclear power is the production of radioactive waste. Nuclear power plants generate various types of radioactive waste, including spent fuel rods, which remain highly radioactive for thousands of years. Proper disposal and management of this waste is crucial to prevent environmental contamination and protect human health. Engineers play a vital role in developing safe and efficient methods for waste storage and disposal.

Another environmental impact of nuclear power is the potential for accidents and the release of radioactive materials. Although rare, accidents such as the Chernobyl and Fukushima disasters have demonstrated the catastrophic consequences of nuclear accidents. Engineers must continually work to improve the safety features and emergency response capabilities of nuclear power plants to minimize the risk of accidents and their environmental impacts.

Furthermore, the mining and processing of uranium, the fuel used in nuclear reactors, can have adverse environmental effects. Uranium mining can cause habitat destruction, water pollution, and the release of radioactive particles into the environment. Engineers need to explore innovative and sustainable methods for uranium extraction and processing to mitigate these impacts.

In conclusion, understanding the environmental impacts of nuclear power is crucial for engineers in the field of energy engineering. While nuclear power offers several advantages, such as low greenhouse gas emissions and a reliable power supply, it also presents challenges, including the management of radioactive waste and the potential for accidents. By addressing these challenges and working towards sustainable solutions, engineers can facilitate the responsible and safe use of nuclear power to meet the world's growing energy demands while minimizing its environmental impacts.

Nuclear Waste Management and Decommissioning

As engineers in the field of energy engineering, it is crucial to possess a comprehensive understanding of nuclear waste management and decommissioning. This subchapter will delve into the complexities and challenges associated with handling nuclear waste, as well as the processes involved in decommissioning nuclear power systems.

Nuclear waste management is a critical aspect of ensuring the safe and sustainable operation of nuclear power plants. The process begins with the categorization of waste into high-level and low-level waste. High-level waste, primarily consisting of spent fuel rods, is highly radioactive and requires careful handling and containment. On the other hand, low-level waste refers to materials that have been contaminated but have lower levels of radioactivity.

One of the primary concerns with nuclear waste management is the long-term storage of high-level waste. While the ideal solution is permanent disposal, it is still a topic of debate and research. Engineers play a vital role in designing and implementing storage facilities such as deep geological repositories or interim storage facilities. These designs must prioritize containment, preventing any leakage or release of radioactive materials.

Decommissioning nuclear power systems is another crucial aspect that engineers in the field of energy engineering must understand. Nuclear power plants have a limited operational lifespan, and decommissioning involves the safe dismantling and disposal of these facilities once they are no longer in use. This process ensures that the site is returned to a safe and environmentally friendly state.

Decommissioning involves several steps, including decontamination, dismantling of equipment, waste management, and site restoration. Engineers must be well-versed in the techniques and technologies required for decommissioning, ensuring that all radioactive materials are handled safely and disposed of properly.

Furthermore, engineers must also consider the economic and social aspects of nuclear waste management and decommissioning. Factors such as cost estimation, public perception, and regulatory compliance play a significant role in decision-making.

In conclusion, nuclear waste management and decommissioning are crucial topics for engineers in the field of energy engineering. With the increasing demand for sustainable and clean energy sources, a thorough understanding of these processes is necessary to ensure the safe and responsible operation of nuclear power systems. By implementing effective waste management strategies and employing appropriate decommissioning techniques, engineers can contribute to a sustainable and secure future in the energy sector.

As engineers specializing in energy engineering, it is crucial to have a comprehensive understanding of nuclear waste management and decommissioning in order to ensure the safe and efficient operation of nuclear power systems. This subchapter aims to provide you with a thorough overview of these critical aspects of nuclear power systems.

Nuclear waste management is a complex process that involves the collection, treatment, storage, and disposal of radioactive waste generated from nuclear power plants. It is vital to handle and store nuclear waste properly to minimize the potential harm to both humans and the environment. This subchapter will delve into the various strategies and technologies employed in nuclear waste management, including the classification of waste, its transportation, and the long-term storage solutions such as deep geological repositories.

Decommissioning refers to the process of safely shutting down and dismantling nuclear facilities once their operational life has ended. It is a highly regulated and carefully planned process that requires expertise in engineering, radiological protection, and environmental management. This subchapter will explore the different stages of decommissioning, including the initial planning, characterization of the facility, dismantling techniques, waste management during decommissioning, and the final site restoration.

Additionally, the subchapter will discuss the challenges and concerns associated with nuclear waste management and decommissioning. These include the long-term storage of radioactive waste, potential risks to human health and the environment, and the financial implications of decommissioning. It is essential for engineers to be aware of these challenges and work towards developing sustainable and safe solutions.

Throughout this subchapter, we will emphasize the importance of adhering to strict regulations and international best practices in nuclear waste management and decommissioning. We will also highlight the research and development efforts aimed at finding innovative solutions to improve the efficiency and safety of these processes.

By gaining a thorough understanding of nuclear waste management and decommissioning, engineers in the field of energy engineering can play a crucial role in ensuring the continued growth and sustainability of nuclear power systems. This subchapter aims to equip you with the knowledge and tools necessary to tackle the challenges and responsibilities associated with these critical aspects of the industry.

Future Perspectives on Nuclear Energy

As engineers, it is crucial to stay informed about the latest advancements and future perspectives in the field of nuclear energy. The continuous growth in energy demand and the urgent need to reduce greenhouse gas emissions have made nuclear power an essential component of the global energy portfolio. This subchapter aims to provide an overview of the future perspectives on nuclear energy, specifically tailored to engineers specializing in energy engineering.

1. Advanced Reactor Technologies: The future of nuclear energy lies in advanced reactor technologies that offer enhanced safety, efficiency, and sustainability. Engineers in the field of energy engineering must familiarize themselves with innovative reactor designs such as small modular reactors (SMRs), advanced fast reactors, and thorium-based reactors. These technologies promise improved safety features, reduced waste generation, and better utilization of nuclear fuel.

2. Generation IV Reactors: Generation IV reactors represent the next phase of nuclear power systems. These reactors are designed to meet the highest safety standards while maximizing energy output. Engineers need to stay updated on Generation IV reactor designs, including the sodium-cooled fast reactor, the molten salt reactor, and the high-temperature gas-cooled reactor. Understanding their unique features and potential applications will be crucial for future energy engineers.

3.   Fusion   Power: Fusion power offers immense potential for clean and abundant energy generation. Engineers specializing in energy engineering should keep a close eye on the development of fusion reactors, such as the International Thermonuclear Experimental Reactor (ITER) project. Fusion power has the advantage of virtually unlimited fuel availability, no long-lived nuclear waste, and inherent safety features. Familiarity with fusion reactor designs and their integration into the energy grid will be vital for engineers in the future.

4.   Nuclear   Energy   and   Renewable   Integration: The future of energy lies in a diverse mix of energy sources. Engineers must explore ways to integrate nuclear energy with renewable sources such as wind and solar power. Developing smart grid technologies, energy storage solutions, and advanced control systems will be essential to achieve a reliable and sustainable energy system.

In conclusion, engineers specializing in energy engineering must be aware of the future perspectives on nuclear energy. Advanced reactor technologies, Generation IV reactors, fusion power, and the integration of nuclear energy with renewables are key areas to focus on. By understanding and embracing these advancements, engineers can contribute to the development of safe, sustainable, and efficient nuclear power systems.

The future of nuclear energy holds great promise as a viable and sustainable solution to the world's growing energy needs. As engineers in the field of energy engineering, it is crucial for us to stay ahead of the curve and understand the potential future perspectives on nuclear energy. This subchapter aims to provide an overview of the exciting developments and challenges that lie ahead for nuclear power systems.

One of the key future perspectives on nuclear energy is the advancement of advanced reactor technologies. Traditional nuclear power plants have been based on light water reactors (LWRs), which have served us well for decades. However, emerging technologies such as small modular reactors (SMRs) and Generation IV reactors offer improved safety features, increased efficiency, and reduced waste production. These advanced reactors have the potential to revolutionize the nuclear energy landscape and provide a more sustainable and reliable source of power.

Another future perspective is the integration of nuclear power with renewable energy sources. As the world moves towards a low-carbon future, it is essential to combine the benefits of nuclear energy with other renewable sources such as solar and wind. This integration can provide a more stable and consistent power supply, as nuclear energy can act as a reliable baseload source while renewable sources fluctuate. Additionally, advancements in energy storage technologies will play a crucial role in facilitating this integration.

Furthermore, future perspectives on nuclear energy also involve addressing concerns related to waste management and proliferation risks. Research and development efforts are focused on developing advanced fuel cycles and recycling technologies to minimize waste and extract more energy from spent fuel. Additionally, the implementation of stringent safeguards and non-proliferation measures will ensure the peaceful use of nuclear energy.

Lastly, the future of nuclear energy also relies on public perception and acceptance. As engineers, it is vital for us to communicate the benefits and safety features of nuclear power to the public. Creating awareness about the role of nuclear energy in reducing greenhouse gas emissions and meeting the growing energy demands can help overcome the stigma associated with nuclear power.

In conclusion, the future perspectives on nuclear energy are promising for engineers in the field of energy engineering. Advanced reactor technologies, integration with renewable sources, waste management improvements, and public acceptance are some of the key areas that will shape the future of nuclear power systems. By staying informed and actively participating in research and development efforts, engineers can contribute to the advancement of nuclear energy and its crucial role in achieving a sustainable energy future.

Chapter 9: Advanced Nuclear Power Technologies

Generation IV Reactor Concepts

In recent years, the field of nuclear energy has witnessed significant advancements in reactor technology. As engineers working in the niche of energy engineering, it is crucial to stay updated with the latest developments in this field to effectively harness the immense potential of nuclear power. This subchapter will provide an overview of Generation IV reactor concepts, which represent the cutting-edge of nuclear technology.

Generation IV reactors are a new breed of nuclear power systems that offer enhanced safety, sustainability, and efficiency compared to their predecessors. These reactors are designed to address the challenges faced by earlier generations, such as the management of nuclear waste, reducing the risk of accidents, and improving overall performance. By understanding the key concepts behind Generation IV reactors, engineers can contribute to the advancement of sustainable and safe energy solutions.

One of the prominent concepts within Generation IV reactors is the use of advanced fuels. These fuels, such as liquid metal, gas, or molten salt, offer superior thermal properties and increased safety margins. They also enable higher burn-up rates, leading to improved fuel utilization and reduced waste generation. Engineers need to grasp the intricacies of these advanced fuels to design reactors that can harness their benefits.

Another significant aspect of Generation IV reactors is their inherent safety features. These reactors incorporate passive cooling systems that rely on natural forces, such as convection and gravity, to dissipate heat, eliminating the need for active mechanical systems. This design approach minimizes the risk of accidents and ensures that even in the event of a failure, the reactor can safely shut down without the need for external intervention.

Furthermore, Generation IV reactors aim to tackle the challenge of nuclear waste management. Many concepts under this category focus on recycling and transmutation of nuclear waste, transforming long-lived radioactive isotopes into shorter-lived ones or reducing their volume. Engineers must explore these innovative approaches to contribute to the development of sustainable solutions for nuclear waste disposal.

In conclusion, Generation IV reactor concepts represent the future of nuclear energy. As engineers specializing in energy engineering, it is essential to familiarize ourselves with these advancements to drive the industry forward. By embracing advanced fuels, inherent safety features, and innovative waste management techniques, engineers can contribute to the development of sustainable and efficient nuclear power systems. This knowledge will empower us to design and implement solutions that can meet the world's growing energy demands while minimizing environmental impact.

In today's world, the demand for clean and sustainable energy sources has never been greater. As engineers in the field of energy engineering, it is our responsibility to explore and develop innovative solutions to meet this demand. One such solution that holds great promise is the concept of Generation IV reactors.

Generation IV reactors represent the next generation of nuclear power systems, designed to be safer, more efficient, and more sustainable than previous generations. These reactors utilize advanced technologies and materials to address the challenges faced by earlier designs, such as waste management, proliferation resistance, and operational safety.

One of the key characteristics of Generation IV reactors is their ability to efficiently utilize nuclear fuel. These reactors make use of advanced fuel cycles, such as the closed fuel cycle, which allows for the recycling and reprocessing of spent fuel. This not only reduces the amount of waste generated but also enhances the overall fuel utilization, maximizing the energy output from a given amount of fuel.

Another significant feature of Generation IV reactors is their enhanced safety features. These reactors incorporate passive safety systems that rely on natural processes and phenomena, ensuring safe shutdown and cooling even in the absence of active human intervention or power supply. This inherent safety greatly reduces the risk of accidents and the potential release of radioactive materials.

Furthermore, Generation IV reactor concepts emphasize sustainability by utilizing alternative energy sources and reducing environmental impact. Some concepts, such as the high-temperature gas-cooled reactor, can be coupled with other energy systems, such as hydrogen production or desalination plants, providing clean and sustainable solutions for various industries and applications.

It is important for engineers in the field of energy engineering to stay informed and up-to-date with the latest advancements in Generation IV reactor concepts. These reactors have the potential to revolutionize the way we generate and utilize nuclear power, offering a safer, more efficient, and environmentally friendly alternative to traditional energy sources.

In conclusion, Generation IV reactor concepts are at the forefront of nuclear power system development. As engineers, it is our duty to explore and implement these advanced technologies to meet the growing global demand for clean and sustainable energy. By understanding the principles and applications of Generation IV reactors, we can contribute to a brighter and more sustainable future for generations to come.

Small Modular Reactors (SMRs)

Small Modular Reactors (SMRs) are gaining significant attention in the field of energy engineering due to their potential to revolutionize the nuclear power industry. These reactors are characterized by their small size and flexible deployment options, offering several advantages over traditional large-scale nuclear power plants.

One of the key benefits of SMRs is their inherent safety features. Their small size allows for a simplified design, reducing the risk of accidents and the potential release of radioactive materials. Additionally, many SMR designs incorporate passive safety systems that rely on natural forces such as gravity and convection, minimizing the need for active intervention during emergencies.

The modular nature of SMRs enables easy and cost-effective deployment. These reactors can be manufactured in factories, transported to the site, and installed without the need for extensive on-site construction. This significantly reduces both the construction time and capital costs associated with nuclear power plants, making SMRs an attractive option for regions with limited resources or grid constraints.

Another advantage of SMRs is their ability to provide flexible power output. Unlike large nuclear power plants that typically have fixed power outputs, SMRs can be deployed in clusters to meet specific energy demands. This scalability makes them suitable for various applications, including remote communities, industrial facilities, and even as a complement to renewable energy sources, providing consistent baseload power.

Furthermore, SMRs offer enhanced versatility in terms of fuel options. Some designs utilize advanced fuel types that increase fuel efficiency and reduce waste generation. Additionally, SMRs can also utilize spent fuel from traditional reactors, contributing to the reduction of nuclear waste stockpiles and increasing the sustainability of the nuclear fuel cycle.

While there are numerous benefits associated with SMRs, there are also challenges that need to be addressed. These include regulatory hurdles, public perception, and the need for standardized designs. However, ongoing research and development efforts are actively addressing these challenges, and several SMR designs are currently under construction or in advanced stages of development.

In conclusion, Small Modular Reactors (SMRs) hold great promise for the future of nuclear power systems. Their small size, inherent safety features, flexible deployment options, and versatile fuel choices make them an attractive option for engineers in the field of energy engineering. As the demand for clean, reliable, and sustainable energy continues to rise, SMRs have the potential to play a significant role in meeting these requirements while ensuring a safe and resilient energy future.

As engineers in the field of energy engineering, it is crucial to stay up-to-date with the latest advancements in nuclear power systems. One such development that has gained significant attention in recent years is the emergence of Small Modular Reactors (SMRs). In this subchapter, we will delve into the concept of SMRs and explore their principles and applications.

SMRs are nuclear reactors that are significantly smaller in size compared to traditional nuclear power plants. They typically have a power output of less than 300 megawatts, making them highly suitable for various applications, including remote areas, military bases, and small grids. Unlike their larger counterparts, SMRs can be manufactured in factories and transported to their deployment sites, resulting in reduced construction time and costs.

The principles behind SMRs mirror those of conventional nuclear reactors. These reactors use nuclear fission to generate heat, which is then used to produce steam to drive turbines and generate electricity. However, SMRs offer enhanced safety features and improved efficiency. Many SMRs utilize passive safety systems that rely on natural phenomena, such as gravity and convection, to shut down the reactor without the need for external power or human intervention. This inherent safety makes SMRs an attractive option for engineers working in the energy sector.

The applications of SMRs are diverse and hold immense potential for addressing the energy needs of various niches. For instance, in remote areas with limited access to a reliable power grid, SMRs can provide a stable and sustainable source of electricity. Additionally, SMRs can be used in combination with renewable energy sources, such as wind and solar, to create hybrid power systems that ensure uninterrupted power supply. Furthermore, SMRs can play a vital role in supporting the electrification of transportation by providing clean energy for electric vehicle charging stations.

In conclusion, Small Modular Reactors (SMRs) represent an exciting development in the field of nuclear power systems. With their compact size, enhanced safety features, and diverse applications, SMRs have the potential to revolutionize the energy industry. As engineers specializing in energy engineering, it is essential to stay informed about the latest advancements in SMRs, as they may hold the key to addressing the energy needs of various niches and contributing to a sustainable future.

Fusion Energy and Tokamaks

In the pursuit of a sustainable and clean energy future, engineers have turned their attention to fusion energy as a potential solution. Fusion energy, often referred to as the "holy grail" of energy production, has the potential to provide an almost limitless supply of clean and safe power. One of the most promising technologies in this field is the tokamak.

A tokamak is a device that uses a magnetic field to confine a plasma, which is a hot and electrically charged gas consisting of ions and electrons. The goal of a tokamak is to heat the plasma to extremely high temperatures, typically in excess of 100 million degrees Celsius, and sustain the conditions required for fusion reactions to occur.

The concept behind a tokamak is based on the idea of using magnetic fields to confine and control the plasma. By creating a toroidal (doughnut-shaped) magnetic field, the plasma can be kept away from the walls of the device, preventing it from cooling down and losing its energy. This magnetic confinement is crucial for maintaining the high temperatures required for fusion reactions to take place.

One of the key challenges in tokamak design is achieving a state of "plasma confinement" known as "plasma equilibrium." This refers to the ability of the magnetic field to keep the plasma stable and prevent it from escaping or losing its energy. Engineers have made significant progress in this area, with advancements in magnetic field design and control systems.

The tokamak design also relies on a combination of heating methods to reach the necessary temperatures for fusion. One commonly used method is known as "ohmic heating," which involves passing an electric current through the plasma to heat it. Additional heating techniques, such as radiofrequency heating and neutral beam injection, are also employed to achieve the desired plasma conditions.

While significant progress has been made in the field of tokamaks, there are still many technical challenges that need to be overcome before fusion energy can become a practical reality. These challenges include finding suitable materials to withstand the extreme conditions inside the tokamak, improving plasma confinement and stability, and developing efficient methods for extracting energy from the fusion reactions.

However, despite these challenges, the potential benefits of fusion energy and tokamaks are undeniable. Fusion energy offers a secure and virtually limitless supply of power, with no greenhouse gas emissions or long-lived radioactive waste. It has the potential to revolutionize the energy industry and provide a sustainable solution for future generations.

As engineers in the niche of energy engineering, it is crucial to stay updated on the latest advancements in fusion energy and tokamak technology. By understanding the principles and applications of these systems, engineers can contribute to the development of a viable and clean energy source that can help meet the growing global energy demand.

In recent years, the quest for sustainable and clean energy sources has become increasingly urgent. Among the various options available, fusion energy has emerged as a promising solution to meet the world's growing energy demands while minimizing environmental impact. This subchapter focuses on fusion energy and tokamaks, which are key components in harnessing this revolutionary power source.

Fusion energy, often referred to as the "holy grail" of energy production, is the process by which two light atomic nuclei, typically isotopes of hydrogen, combine to form a heavier nucleus, releasing an enormous amount of energy in the process. This reaction is the same process that powers the sun and other stars, making it an abundant and virtually limitless source of energy.

Tokamaks, a type of magnetic confinement device, are at the forefront of fusion energy research. The word "Tokamak" is derived from the Russian acronym for "toroidal chamber with magnetic coils." These devices confine a hot plasma of ionized gases, typically deuterium and tritium, using powerful magnetic fields to achieve the necessary conditions for fusion reactions to occur.

This subchapter delves into the principles behind tokamaks and their operational characteristics. It explores the key components of a tokamak, such as the toroidal field coils, poloidal field coils, and plasma heating systems. Additionally, it discusses the challenges associated with confining and controlling the plasma, including plasma instabilities and disruptions, and the strategies employed to mitigate these issues.

Furthermore, this subchapter provides an overview of current research and development efforts in the field of fusion energy, highlighting the advancements made in plasma physics, fusion reactor designs, and materials science. It also discusses the potential benefits and challenges of fusion energy, including its high energy density, low environmental impact, and the need for advanced technologies and international collaboration to achieve commercial viability.

Engineers and professionals in the niche of energy engineering will find this subchapter valuable in gaining a comprehensive understanding of fusion energy and tokamaks. It equips them with the knowledge to contribute to the ongoing research and development in fusion energy technologies and explore the potential applications of fusion energy in addressing the world's energy challenges.

Overall, fusion energy and tokamaks present a captivating and promising field of study that holds the potential to revolutionize the way we produce and consume energy. This subchapter serves as an essential resource for engineers seeking to stay at the forefront of this groundbreaking technology and contribute to a sustainable and clean energy future.

Nuclear Power in Space Exploration and Propulsion

As engineers in the field of energy engineering, it is essential to explore new frontiers and push the boundaries of technology. One such frontier is space exploration, where the demand for reliable and efficient power systems is crucial. In this subchapter, we will delve into the fascinating world of nuclear power and its applications in space exploration and propulsion.

Space missions involve extended periods in space, often reaching distances where solar power becomes impractical. This is where nuclear power systems come into play. Nuclear power offers a compact and reliable source of energy that can be utilized in space exploration missions. It provides a continuous power supply, allowing spacecraft to operate even in the darkest regions of space.

One of the most significant applications of nuclear power in space exploration is radioisotope thermoelectric generators (RTGs). RTGs use the heat generated from the natural decay of radioactive isotopes, such as plutonium-238, to produce electricity. This technology has been used in various space missions, including the Voyager spacecraft, which has been exploring the outer reaches of our solar system for over four decades.

Another promising application is nuclear propulsion. Nuclear propulsion systems can provide far greater thrust and efficiency than conventional chemical propulsion systems. By utilizing nuclear reactors to heat propellant, spacecraft can achieve higher velocities and significantly reduce travel times for long-distance missions. This technology has the potential to revolutionize space exploration, enabling faster and more efficient interplanetary travel.

However, the implementation of nuclear power in space exploration and propulsion comes with its own set of challenges. Safety and containment of radioactive materials are of utmost concern. Engineers must design robust and fail-safe systems to prevent any potential accidents or environmental contamination.

Moreover, the regulatory and political landscape surrounding nuclear power in space must be carefully navigated. International agreements and protocols need to be established to ensure the responsible use of nuclear technology in space missions.

In conclusion, nuclear power holds immense potential for space exploration and propulsion. As engineers in the field of energy engineering, it is our responsibility to further develop and refine these technologies. By overcoming the technical, safety, and regulatory challenges, we can unlock new frontiers in space exploration and pave the way for humanity's future among the stars.

Introduction

As humanity ventures further into space exploration, the need for advanced propulsion systems becomes increasingly evident. Traditional chemical rockets have limitations in terms of speed and efficiency. In this subchapter, we will explore the role of nuclear power in space exploration and propulsion, specifically addressing its applications, benefits, and challenges. This content is addressed to engineers specializing in energy engineering.

Applications of Nuclear Power in Space Exploration

Nuclear power offers several advantages over conventional propulsion systems. One of the most significant applications is nuclear thermal propulsion (NTP), which utilizes the heat generated by nuclear reactions to heat a propellant, such as liquid hydrogen, resulting in higher exhaust velocities and improved efficiency. NTP systems have the potential to reduce travel times and enable more ambitious missions, such as crewed missions to Mars.

Another promising application is nuclear electric propulsion (NEP). NEP systems utilize nuclear reactors to generate electricity, which in turn powers electric thrusters. These thrusters, such as ion engines or Hall-effect thrusters, provide low thrust but high specific impulse, enabling efficient long-duration space missions. NEP systems have already been employed in various deep space missions, including NASA's Dawn spacecraft.

Benefits and Challenges

Nuclear power offers numerous benefits for space exploration. The high energy density of nuclear fuels allows for compact power systems, which is crucial for spacecraft where weight and size are critical factors. Additionally, nuclear power enables longer-duration missions by providing a continuous power source, eliminating the need for solar panels in regions with limited sunlight, such as outer planets or interstellar space.

However, nuclear power in space also presents challenges. Safety is a primary concern, as any malfunction or accident could have severe consequences. Engineers must design robust and fail-safe systems to prevent potential radiation leaks or reactor meltdowns. Moreover, the disposal of nuclear waste generated by these systems must be addressed to ensure minimal environmental impact.

Conclusion

Nuclear power has the potential to revolutionize space exploration and propulsion systems. Its applications in nuclear thermal propulsion and nuclear electric propulsion offer improved efficiency, reduced travel times, and the ability to undertake ambitious missions. While the benefits are significant, engineers specializing in energy engineering must address the challenges associated with safety and waste disposal to ensure the responsible and sustainable use of nuclear power in space. By overcoming these challenges, we can unlock the full potential of nuclear power, propelling humanity further into the cosmos.

Chapter 10: Case Studies and Real-World Applications

Nuclear Power Plants Around the World

Introduction:

In this subchapter, we will explore the fascinating world of nuclear power plants around the globe. As engineers in the field of energy engineering, it is crucial to understand the current state of nuclear power and its applications. From advanced technologies to safety measures, this section will shed light on the different nuclear power plants operating worldwide.

Global   Overview:
Nuclear power plants have become a significant source of electricity generation across the globe. As of [current year], there are [number] nuclear power plants in operation, with a combined capacity exceeding [capacity] gigawatts. These plants are spread across various countries, including the United States, France, China, Russia, and Japan, among others.

Advanced   Reactor   Technologies:
One of the key aspects of nuclear power plants is the use of advanced reactor technologies. Engineers in the field of energy engineering are continuously working on innovative designs to enhance safety, efficiency, and sustainability. Breeder reactors, high-temperature gas-cooled reactors, and molten salt reactors are examples of advanced technologies being explored in different countries.

Safety   Measures:
Safety is of utmost importance in the nuclear power industry. Engineers play a critical role in designing and implementing safety measures to prevent accidents and mitigate potential risks. This subchapter will discuss the various safety protocols and features in place at nuclear power plants worldwide, including redundant safety systems, containment structures, and emergency response plans.

International   Cooperation:
The field of energy engineering is marked by international collaboration and knowledge-sharing. Organizations such as the International Atomic Energy Agency (IAEA) facilitate cooperation among countries to ensure the safe operation of nuclear power plants. This subchapter will highlight the importance of international cooperation in the nuclear power industry and the role engineers play in fostering global collaboration.

Current   Challenges   and   Future   Prospects:
While nuclear power plants offer numerous benefits, they also face challenges. These include waste disposal, public perception, and the need for continuous technological advancements. As engineers, it is crucial to be aware of these challenges and work towards finding sustainable solutions. This subchapter will discuss these challenges and explore the future prospects of nuclear power, including the potential for fusion reactors and small modular reactors.

Conclusion:
In conclusion, nuclear power plants are a vital component of the global energy mix. As engineers specializing in energy engineering, it is essential to stay updated with the advancements, safety measures,

and international cooperation in the field of nuclear power. This subchapter aims to provide insights into the current state of nuclear power plants around the world, emphasizing the role of engineers in shaping the future of this industry.

Introduction:

Nuclear Power Systems: Principles and Applications for Engineers is a comprehensive guide that aims to provide engineers specializing in energy engineering with a thorough understanding of nuclear power plants around the world. This subchapter explores the global landscape of nuclear power plants, highlighting their significance in meeting the growing energy demands while ensuring sustainable development.

Global Overview of Nuclear Power Plants:

Nuclear power plants are a critical component of the global energy mix, providing a reliable and low-carbon source of electricity. As of [current year], there are [number] nuclear power plants operating in [number of countries] countries worldwide. These plants collectively generate an impressive [number] gigawatts of electricity, supplying a significant portion of the global energy demand.

Regional Variations and Trends:

Nuclear power generation varies across different regions of the world. While some countries have heavily invested in nuclear energy, others have chosen alternative sources. Europe, for instance, has a higher concentration of nuclear power plants, with countries like France, the United Kingdom, and Germany leading the way. In contrast, North

America, Asia, and the Middle East have seen significant growth in the construction of new nuclear power plants in recent years.

Safety Measures and Regulations:

With concerns surrounding nuclear safety, stringent regulations and safety measures have been implemented globally. These measures include the use of advanced reactor designs, rigorous safety protocols, and comprehensive emergency preparedness plans. Engineers play a crucial role in ensuring the safe and efficient operation of nuclear power plants, adhering to international guidelines and continuously improving safety practices.

Advantages of Nuclear Power Plants:

Nuclear power plants offer several advantages, making them an attractive option for meeting the world's energy needs. They produce large amounts of electricity with comparatively low greenhouse gas emissions, contributing to mitigating climate change. Additionally, nuclear power plants provide a reliable and continuous energy supply, reducing dependence on fossil fuels and enhancing energy security.

Challenges and the Future of Nuclear Power Plants:

Despite the advantages, nuclear power plants face challenges related to public perception, waste management, and decommissioning. However, ongoing research and development efforts are focused on addressing these challenges. The future of nuclear power plants lies in the development of advanced reactor technologies, such as small modular reactors and fusion reactors, which offer improved safety, waste reduction, and enhanced efficiency.

Conclusion:

The global landscape of nuclear power plants is continuously evolving, with engineers playing a vital role in ensuring their safe and efficient operation. This subchapter aimed to provide energy engineering professionals with an overview of nuclear power plants worldwide, highlighting their importance, safety measures, advantages, and future prospects. As engineers, it is essential to stay updated with the latest advancements in nuclear power systems to contribute to a sustainable and secure energy future.

Nuclear Accidents and Lessons Learned

In the field of energy engineering, the topic of nuclear accidents is of paramount importance. It is crucial for engineers to understand the history, causes, and consequences of such accidents to continuously improve the safety and reliability of nuclear power systems. This subchapter aims to explore the most significant nuclear accidents and the valuable lessons learned from them.

One of the most notable nuclear accidents in history occurred at the Chernobyl Nuclear Power Plant in 1986. The catastrophic explosion and subsequent release of radioactive materials highlighted the importance of reactor design and emergency response protocols. Engineers learned that flawed reactor design, inadequate safety measures, and a lack of proper training and communication among operators were contributing factors to the incident. As a result, significant improvements in reactor design and safety protocols were implemented worldwide.

Another well-known nuclear accident took place at the Fukushima Daiichi Nuclear Power Plant in 2011, following a severe earthquake and tsunami. This incident revealed vulnerabilities in the face of natural disasters and the importance of disaster preparedness. Engineers learned the significance of robust containment structures, backup power systems, and emergency cooling mechanisms to prevent or mitigate accidents. The lessons from Fukushima prompted the revision of safety regulations, improved emergency response plans, and enhanced seismic design criteria for nuclear power plants globally.

Furthermore, engineers have gained valuable insights from smaller-scale accidents and incidents, such as Three Mile Island in 1979 and the SL-1 reactor in 1961. These incidents highlighted the significance of human factors, operator training, and the importance of effective communication in preventing accidents.

Overall, the lessons learned from nuclear accidents have shaped the development of nuclear power systems and safety practices. Engineers have continuously worked towards enhancing reactor design, strengthening safety features, improving emergency response plans, and ensuring effective operator training. Ongoing research and development have led to the implementation of passive safety systems, advanced reactor designs, and more robust emergency preparedness measures.

As engineers engaged in energy engineering, it is crucial to understand the history and lessons learned from nuclear accidents. This knowledge allows for the continuous improvement of nuclear power systems, ensuring their safety, reliability, and contribution to sustainable energy production. By applying these lessons, engineers can foster a safer and more resilient nuclear industry, providing clean and efficient energy for the future.

As engineers in the field of Energy Engineering, it is crucial for us to understand the potential risks and consequences associated with nuclear power systems. Nuclear accidents have occurred throughout history, and although they are rare, they have had significant impacts on both human life and the environment. This subchapter aims to explore some of the most notable nuclear accidents and the valuable lessons we have learned from them.

One of the most well-known nuclear accidents took place in 1986 at the Chernobyl Nuclear Power Plant in Ukraine. This catastrophic event resulted in a steam explosion and a subsequent fire, releasing a large amount of radioactive material into the atmosphere. The accident had devastating consequences, including immediate deaths and long-term health effects on the surrounding population. Engineers learned valuable lessons from this incident, such as the importance of proper reactor design, safety protocols, and operator training. As a result, significant improvements were made in reactor safety systems and emergency preparedness.

Another notable accident occurred in 1979 at the Three Mile Island Nuclear Generating Station in the United States. A partial meltdown of the reactor core led to the release of a small amount of radioactive gases. While this incident did not result in any immediate casualties, it highlighted the importance of effective communication and transparent information sharing during a nuclear crisis. Engineers recognized the need for clear and accurate communication with the public, as well as the significance of robust safety measures and redundant systems to prevent accidents.

The Fukushima Daiichi nuclear disaster in 2011, triggered by a massive earthquake and tsunami in Japan, further emphasized the importance of disaster preparedness and the need for continuous improvements in reactor safety. Engineers learned valuable lessons from this incident, such as the necessity of locating critical safety systems at higher elevations to protect them from flooding and the importance of adequate backup power sources.

These accidents and the lessons learned from them have led to significant advancements in nuclear power system design and operation. Engineers have developed more robust safety measures, improved emergency response protocols, and enhanced training programs for operators. Additionally, there has been a greater emphasis on public engagement and communication in the nuclear industry.

As engineers in the field of Energy Engineering, it is our responsibility to continuously learn from past nuclear accidents and apply these lessons in the design, operation, and maintenance of nuclear power systems. By implementing the best practices and safety measures developed from these incidents, we can ensure the safe and efficient utilization of nuclear energy to meet the world's growing energy demands.

Nuclear Power in Developing Countries

In recent years, the global energy landscape has been undergoing significant transformations, with developing countries playing an increasingly vital role. As these nations strive to meet the growing energy demands of their rapidly expanding populations and industrial sectors, nuclear power has emerged as a viable solution to address their energy needs. This subchapter aims to provide engineers specializing in energy engineering with a comprehensive overview of nuclear power in developing countries, highlighting its principles and applications.

One of the primary reasons why nuclear power has gained traction in developing countries is its ability to generate large amounts of electricity reliably and sustainably. Unlike fossil fuel-based power plants, nuclear power plants do not emit greenhouse gases, making them an attractive option in the fight against climate change. Furthermore, nuclear power plants offer a stable and continuous power supply, which is crucial for countries experiencing rapid economic growth.

However, the adoption of nuclear power in developing countries is not without its challenges. Building and operating a nuclear power plant requires a significant investment of financial and human resources, as well as extensive safety measures and regulations. Developing countries often face constraints in terms of technological expertise, infrastructure, and financial capabilities. Therefore, it is crucial for engineers in the field of energy engineering to understand the specific considerations and strategies for implementing nuclear power in these contexts.

This subchapter will cover various aspects related to nuclear power in developing countries, including technology selection, regulatory frameworks, safety precautions, and waste management. It will delve into the different types of nuclear reactors available and discuss their suitability for specific regions and energy demands. Additionally, the subchapter will explore the potential benefits and drawbacks of nuclear power in developing countries, considering factors such as cost-effectiveness, energy security, and environmental impact.

By equipping engineers specializing in energy engineering with the knowledge and insights necessary to navigate the complexities of nuclear power in developing countries, this subchapter aims to facilitate informed decision-making and promote sustainable energy solutions. As these nations continue to strive for economic development and energy security, nuclear power has the potential to play a crucial role in meeting their growing energy demands while mitigating the adverse effects of climate change.

In recent years, there has been a growing interest in nuclear power as a viable option for meeting the energy demands of developing countries. As engineers specializing in energy engineering, it is imperative for us to understand the potential benefits and challenges associated with the deployment of nuclear power systems in these nations.

Developing countries face unique energy challenges, including limited access to electricity, unreliable power grids, and a growing need for sustainable and affordable energy sources. Nuclear power has the potential to address these issues by providing a reliable and consistent source of electricity that is independent of weather conditions and offers a large-scale power generation capacity.

One significant advantage of nuclear power is its low carbon footprint. With the global push towards reducing greenhouse gas emissions, nuclear energy provides a clean and sustainable alternative to fossil fuel-based power generation. Developing countries, often heavily reliant on coal or oil for electricity, can greatly benefit from the adoption of nuclear power systems, as it enables them to meet their energy needs while reducing their carbon emissions.

Moreover, nuclear power systems can contribute to technological progress and economic development. The construction and operation of nuclear power plants require a highly skilled workforce, providing employment opportunities and fostering the development of technical expertise within the country. Additionally, the establishment of a domestic nuclear industry can stimulate technological advancements and innovation, leading to spin-off industries and economic growth.

However, the deployment of nuclear power systems in developing countries also presents challenges that must be addressed. These challenges include ensuring safety, security, and non-proliferation measures, as well as the establishment of robust regulatory frameworks. Developing countries may lack the necessary infrastructure, expertise, and financial resources to manage and operate nuclear power plants effectively. Collaboration with international organizations and experienced nuclear countries is crucial in providing support and guidance to ensure the safe and responsible implementation of nuclear power systems.

In conclusion, nuclear power has the potential to play a significant role in meeting the energy demands of developing countries. As engineers specializing in energy engineering, it is our responsibility to explore and evaluate the benefits and challenges associated with nuclear power deployment in these nations. By considering the unique energy needs, environmental concerns, and socio-economic factors, we can contribute to the development of sustainable and resilient energy systems that will foster the growth and prosperity of developing countries.

Future Trends and Challenges in Nuclear Energy Engineering

As engineers in the field of energy engineering, it is crucial to have access to reliable and up-to-date resources that can support your work and expand your knowledge in the field of nuclear power systems. This subchapter titled "References" aims to provide you with a comprehensive list of recommended resources that will serve as valuable references throughout your professional journey.

1. "Nuclear Energy: An Introduction" by Raymond Murray and Keith E. Holbert: This book offers a comprehensive overview of the fundamental principles and applications of nuclear energy. It covers topics such as reactor types, fuel cycles, safety, and environmental impacts, making it an essential reference for engineers in the energy field.

2. "Nuclear Power Plant Systems and Equipment" by Philip Kiameh: This book provides detailed insights into the design and operation of nuclear power plants. It covers various aspects, including reactor systems, components, thermal-hydraulics, safety systems, and maintenance, serving as an excellent reference for engineers involved in the practical aspects of nuclear power systems.

3. "Nuclear Reactor Analysis" by James J. Duderstadt and Louis J. Hamilton: This comprehensive textbook covers the theory, design, and analysis of nuclear reactors. It delves into reactor physics, neutron diffusion, fuel cycle analysis, and core design principles, providing engineers with a solid foundation in nuclear reactor analysis.

4. "Introduction to Nuclear Engineering" by John R. Lamarsh and Anthony J. Baratta: This textbook offers a comprehensive introduction to nuclear engineering principles and applications. It covers topics such as nuclear reactions, radiation detection and measurement, reactor safety, and nuclear fuel cycles, making it an invaluable resource for engineers seeking a broad understanding of the field.

5. "Nuclear Energy: Principles, Practices, and Prospects" by David Bodansky: This book provides a balanced overview of the technical, economic, environmental, and social aspects of nuclear energy. It discusses the history, current state, and future prospects of nuclear power, making it a valuable reference for engineers interested in understanding the broader context of nuclear energy.

6. International Atomic Energy Agency (IAEA) Publications: The IAEA publishes a wide range of technical documents, guidelines, and safety standards related to nuclear power systems. These publications cover topics such as reactor technology, safety, waste management, and radiation protection, providing engineers with authoritative and internationally recognized references.

These references, among many others, will serve as valuable resources to broaden your knowledge, enhance your understanding of nuclear power systems, and stay up-to-date with the latest developments in the field. By utilizing these references, you will be equipped with the necessary information to make informed decisions, design efficient systems, and contribute to the advancement of energy engineering.